Pipe and Tube Bending Manual

Pipe and Tube Bending Manual

SECOND EDITION

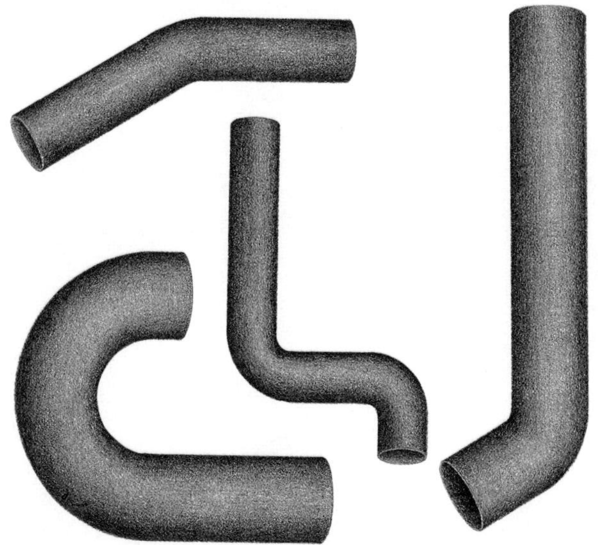

John Gillanders

Published by
Fabricators & Manufacturers Association, International
Rockford, Illinois

ISBN 1-881113-06-X

Printed in the United States of America

CONTENTS

Appendices

Acknowledgments

This book was written with a great deal of help and encouragement from others. I am indebted first to Bert Turner, Chairman of the Board, TIL Group, who despite a recessionary economy had the perspicacity to undertake the construction of an ultra-modern pipe-bending and fabrication facility; to Jim Walsh, of Nichols Construction, who conceived that project and gave me the opportunity of helping him, first as a consultant and then as Director of Technical Development, to put it together and get it under way. In so doing, a great deal of source material was accumulated, much of which provided the information, particularly on certain aspects of newer technology and automation, that is contained in this book. Also, I never could have pulled everything together without the help of Glenda Firor, secretary par excellence, who managed to make sense of all my scribbling and dictation and arrange it into readable form.

I also wish to thank all of the following for their most helpful information, comments, and, in many cases, photographs and technical data: Ray Jursa, DeKay Fabricators; Ken Elrod, Allied Industries; C.W. Eagles, Canadian Industrial Tube Bend; Steve Smith, Texas Pipe Benders; Oscar Shank, Houston Pipe Benders; Mike Ball, Eaton-Leonard Corporation; Tony Granelli, Dow Chemical; Jim Papageorge, Inkamaf Corporation; Geoff Gregson, Gregson Pipework Limited; George Mock, Piping International; Gunter Wilkens, Oxytechnik; Patrick Monroe, Tulsa Tube Bending Company; Charles Russell, Jr., 600 Machinery; Guruswami Ranganathan, Schwarze-Wirtz; Rick Spinelli, Conrac Corporation; Ed Ptack, Wallace Manufacturing Company; Ron Stange, Tools for Bending, Bob Hipley, Associated Piping and Engineering Corporation; and a host of others, too numerous to mention here, with whom I have discussed—and sometimes argued—bending at one time or another over the past several years. I hope they too will take it for granted that they have contributed to this book and that they will accept my deep appreciation and thanks.

PREFACE

Years ago, I suddenly found myself deeply involved in pipe and tube bending without knowing the first thing about it. I had to learn everything the hard way, by trial and error; mostly I learned from my mistakes and, believe me, there were plenty of them. My task, and the task of my associates, was made even harder by the astonishing fact that there existed not even a beginner's guide to which we could refer!

In particular, my sympathies lie with the personnel at International Piping Systems, Ltd., who found themselves having to cope with both cold and induction bending, automated processing, computer-aided design and production and a whole variety of new methods and technologies in a brand new 90,000 square foot plant.

I hope, at last, these people—and any others involved in pipe and tube bending—can now quickly refer to this handy manual for the answers to their innumerable questions.

John Gillanders

1

Introduction

Ever since the first mechanical bending device was patented in the United States more than a hundred years ago, both machinery and the accompanying technology have grown increasingly complex and diversified. Today there are literally dozens of different kinds of machines, ranging from a simple hand bender to a fully computerized numerical control machine capable of producing hundreds of bends an hour.

The largest cold bending machines can handle material up to 20″ in diameter, while induction machines can accommodate pipe up to 70″ in diameter, with wall thicknesses of up to 4″. There is even machinery that makes its own tubing, bends it on several different planes, cuts the finished piece and dispatches it on a conveyor belt, all in a matter of seconds.

Despite these tremendous advances in bending technology, there has never been a truly informative and comprehensive book on the subject—only technical papers presented at conferences and clinics and descriptive literature published by the manufacturers.

This book is intended as a reference manual not only for those presently involved in bending but also for those who are contemplating getting involved, as well as for engineers, designers, architects, manufacturers, fabricators, and a host of others who use bends for one purpose or another. Most people today are totally unfamiliar with bending and are unaware of what can and cannot be done when it comes to putting a bend in a length of pipe or tube. In addition to illustrating and describing some of the many different types of bending and bending machinery, this book focuses on some of the other elements in the production of a bend: tooling, materials, and not least in importance, the human element—the operator himself. Also included are a number of charts and formulas that will be helpful in determining certain bending criteria.

Much of this book is devoted to new and emerging bending technology and focuses attention on such areas as computer-aided design and fabrication, automated production, modular construction, and robotics, where these apply to the fabrication of pipe spools for the hydrocarbon processing and other industries. It is hoped this book will help those who, like myself, have often said, "Where can I find out about . . . ?"

I have tried to avoid getting into highly detailed and complex explanations of how bending is accomplished by the various types and kinds of machinery and to confine matters to a general perspective in each case.

Individual manufacturers will be only too glad to explain the intricacies and advantages of their particular products, and it is not my intention to "sell" any one method of bending or type of machine but rather to explain in simple terms what pipe and tube bending is all about. The field itself is enormous and far too complex to cover in a single book. Even those directly involved in a certain type of bending are often quite unfamiliar with, or even unaware of, the technologies in other areas. This book should help to broaden knowledge of the field.

Although pipe and tube have been lumped together, so to speak, the application of bending in many manufacturing industries has been far wider with tube than with pipe. It is here specifically that I have tried to emphasize the advantages of bending as opposed to welding fabrication, and it is here that we can see the greatest potential for cost-effective fabrication, installation, and end use.

This area in particular has been identified by the Business Round Table Report B-3 of August, 1982 "Construction Technology Needs and Priorities," which comments, "Piping appears to be the most inefficient of all the major areas of physical construction." In its conclusion the report places piping at the head of a list of areas wherein "dramatic economic gains might flow from research and technological improvement. . . ."

It has always bothered me that piping engineers and designers pay so little attention to a technology that could have saved many millions of dollars if it had been addressed more seriously and vigorously in the past.

Both machinery manufacturers and benders know the cost savings that can be achieved by using a bend to turn a corner in piping rather than using a butt weld fitting. Yet we have continued to use conventional welding fabrication in piping design and construction for years, without really researching and developing the full potential of pipe bending technology. We have very few national standards for pipe bending and have never really focused much attention on this area. It is high time we did if we are to recover our former position as a world leader among industrial nations.

Bending Terminology

There are a number of terms and expressions used in bending pipe and tube which may not be familiar to everybody. Terms, such as radius, degree of bend, and plane of bend are fairly common. Others, however, when used by a bender, may bring a blank stare or raised eyebrow. The following explanations should be helpful in this case.

Center to Center—The distance between the theoretical or calculated centers of two adjoining bends on the same plane. Also used for the diametric measurement between the centerlines of two tangent points of a bend, i.e., 180° bend, for which the center-to-center distance will be equal to twice the centerline radius.

Compression—A type of mechanical bending without the use of a mandrel. Also, the forces that thicken the inside (intrados) of the bend.

Degree of Bend (DOB)—This relates strictly to the number of degrees required in a particular bend. A right angle, for instance, would be 90° of bend and a U bend would be 180°. One seldom finds a need for a bend of less than 3° or greater than 180°, since both present additional problems in achieving bends of this nature by the draw bending method. It is, of course, entirely possible to make a bend of 360°, but this is normally done by what is called the roll forming or coil forming process and cannot be done by the draw process under most circumstances.

Distance Between Bends (DBB)—The straight section between the tangent points of two adjoining bends.

Elongation—The amount of stretch undergone by the material fiber during bending.

Extrados—The outside arc of the bend.

ID—The inside diameter of a pipe or tube.

Intrados—The inside arc of the bend.

Moment—Refers to the moment of inertia. The point at which a tube or pipe section starts to change its shape, measured in foot-pounds or ton-inches.

Neutral Axis—The centerline axis of the pipe or tube along the arc of the bend.

Nominal—Usually refers to pipe sizes up to and including 12″ IPS. Also used in reference to wall thickness, generally as a "mean" or average measurement.

OD—The outside diameter of a pipe or tube.

Ovality—The distortion of the cross section of pipe or tube from its normal (round) shape.

Plane of Bend (POB)—The plane of the bend in relation to the axis of the straight section preceding it.

Radius—This is the arc of the bend itself and is usually taken at the centerline of the material or neutral axis. The radius may be of any length compatible in its shortest distance with the bendability, so to speak, of the material itself. Usually a radius is referred to in terms of multiples of the OD of the material, i.e., a 3D (3 × diameter) bend in 2″ OD tube would be a 6″ centerline radius (3 × 2). However, this is misleading when dealing with pipe, as nearly all pipe from 1/8″ to 12″ IPS is nominal in size. Thus, 4″ pipe actually has an outside diameter of 4 1/2″ (see Appendix A) and a 3D bend in 4″ pipe, while formed on a radius of 12″ would actually have a true D factor of less than 3D, in this case 1.33D. This may not seem very important, but when determining the nature and type of tooling required to make a critical tight-radius bend, it can be extremely important. We will discuss this in the chapter on tooling.

Springback—The attempt by pipe or tube material to return to its original configuration where it has not been stressed to its yield point after bending.

Tangent Points—These are two points at which a bend starts and finishes. Every bend regardless of size, radius, or degree of bend has two tangent points.

Tangent—The straight material at each end of the bend is referred to as the tangent and may be of any length. Thus, a bend that has no straight on either end would have zero tangents whereas one with a foot of straight at each end would have 12″ tangents. Again, while zero or very short tangents may be required of the finished bend, in order to make the bend itself there has to be sufficient material at each end during the bending process, any excess material usually being trimmed off afterwards.

Tensile Strength—The point at which material stretched beyond its yield will rupture.

Tension—The stretch forces on the outside (extrados) of the bend.

Wall Thickness—All hollow material, be it pipe or tube, has a certain wall thickness. In pipe it is usually measured in standards of thickness called schedules and in tube it is usually measured in thicknesses of a gauge. For instance, 2″ 14-gauge tube has an outside diameter of 2″ and a wall thickness of .083″; 16-gauge tube has a lesser wall thickness of .065″. As the gauge increases so the wall thickness decreases. Pipe, on the other hand, goes the opposite way, and 2″ Schedule 10 pipe has an outside diameter of 2³/₈″—remember, pipe up to 12″ is in nominal sizes and the Schedule 10 for that size material has a wall thickness of .109″. The same size pipe in Schedule 40 has a wall thickness of .154″ and in double extra heavy (XXH) material a thickness of .436″. Thus, as the schedule increases so does the wall thickness and this will vary from one pipe size to the next, whereas for tube the gauge will remain constant regardless of the OD of the material. The relationship between outside diameter and wall thickness has an important bearing on tooling selection, which is dealt with in Chapter 4.

Wall-Thinning—The amount of reduction from normal or nominal wall thickness to the amount of wall thickness remaining in the extrados of a bend after forming.

Yield Point—The point at which material will deform permanently during bending.

See the Key elements of a bend in Figure 1-1.

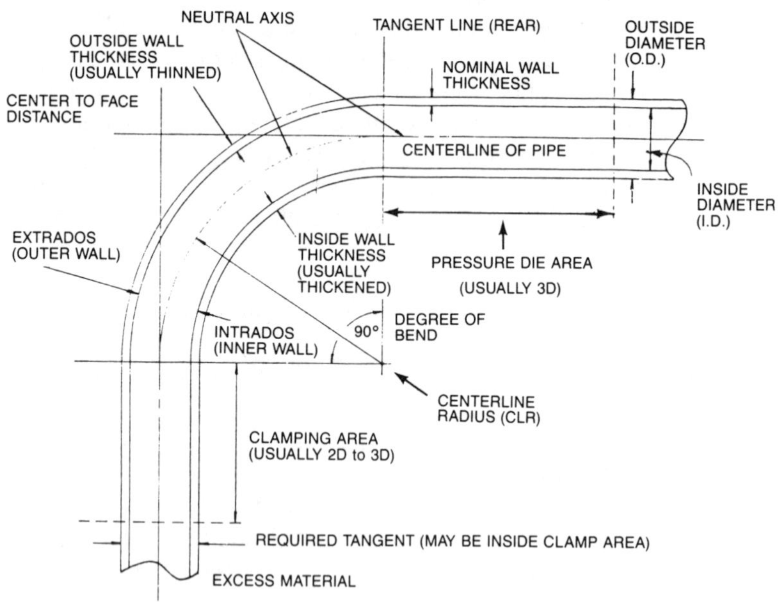

Figure 1-1. Key elements of a bend.

Material, Machine and Tooling Implications

Before starting out to bend a particular piece of pipe or tube, a great many factors must be taken into consideration, all of which will have a significant bearing on the end results. In fact, they will decide whether the workpiece can even be bent and which type or size of machine must be used.

For example, one might assume a piece of 4″ pipe would normally be bent on a 4″ machine. However, all machines have limits and 4″ draw benders for minimum 1½D radii will usually have a maximum capability of bending 4″ Schedule 80 carbon steel pipe at 1½D. They will not bend Schedule 160 pipe on that radius nor will they accommodate stainless steel in Schedule 80 at 1½D. If the radius is not as tight as 1½D and is, say, 5D, it is possible the machine can handle this. But if the machine's maximum radius is less than the minimum achievable radius in the Schedule 160 carbon steel or Schedule 80 stainless steel, the bend will have to be done on a larger machine.

Before anything else, therefore, it is important to know what a machine's limitations are. Once these are determined, one can examine the material and other factors of the piece to be bent.

Most of the pertinent data will be contained in the pipe and tube manufacturers' handbooks. In addition to chemical properties, such details as tensile strength, yield point, section modulus, variations in wall thickness, OD, and ID must all be considered.

For example, A53, A106, and A120 carbon steel pipe have basically the same minimum wall thickness tolerance of not more than 12½% less than the nominal wall. Thus, a piece of 4″ Schedule 40 pipe with a nominal wall thickness of .237″ could, at certain points along its length, have an actual wall thickness of .207375″. If one put a bend in the pipe with the minimum wall thickness in the extrados of the bend area and the bend was thinned out 18%, we would end up with an actual wall thickness of approximately .170″, or about 28% reduction from the nominal.

The OD of the pipe may also vary, and this can affect the bend quality since oversize material will show tooling marks if the tooling itself is very precise. A53 and A120 pipe, for instance, may be as much as ±1% for 2″ and larger material. This means that 12″ pipe can be slightly more than 1/8″ over- or undersized or there can be a variation between the two of more than 1/4″.

For cold bending, involving the use of precision tooling, e.g., mandrel, wiper, clamp, and pressure die, the variations can be troublesome and, if everything is at the limit of permissible tolerance, problems can arise such as having the mandrel jam, which usually results in a bent mandrel rod. With OD tolerances on the plus side, scoring and marking the pipe in the bend and clamp areas is a common result.

For induction bending, however, the problems are not as critical since there is virtually no tooling involved.

Tooling plays a prominent role in cold bending and one must select the tooling required in accordance with certain factors. These are readily determined and will dictate the tooling to be used.

For example, to bend 6″ Schedule 10 stainless steel pipe on a radius of 9″ (1 1/2D), we can determine that it will be necessary to use a close pitch three-ball mandrel and wiper die. The same material will require only a Roberts Regular two-ball mandrel and wiper die for an 18″ radius (3D). For a 30″ radius (5D) only a one-ball or two-ball Roberts Regular mandrel would be required and no wiper die. The formulas are easy to work out and are as follows:

$$\text{Wall factor} = \frac{\text{OD}}{\text{t}} = \frac{6.625}{.134}$$

where OD is the true outside diameter, *not* the nominal size and t is the actual wall thickness.

For 6″ Schedule 10, therefore, the wall factor will be 49.44 or to the closest rounded figure a wall factor of 50.

$$\text{Bend factor} = \frac{\text{R}}{\text{OD}} = \frac{9}{6.625}$$

where R equals centerline radius and OD equals true OD

The bend factor, therefore, will be 1.358D.

This true D in bend factor raises the issue of the real differences between nominal D and true D for all pipe in the 1/8″ to 12″ IPS size range. When determining a particular bending requirement, in addition to the physical dimensions of the pipe or tube in relation to machine design capability, certain other material properties and machine capabilities must also be considered. These are:

1. Yield point
2. Tensile strength
3. Percentage of elongation of the material
4. Springback
5. Torque bending moment of the machine

When the pipe/tube is stretched and/or compressed to the point where it starts to deform and stay deformed, it is considered to have reached its *yield point,* which is expressed in psi (stress in pounds per square inch). If the material is flexed slightly and returns to its original shape, it has not reached the yield point. One can see this simply by lifting one end of a 20′ length of 2″ pipe—it will bend slightly but return to its straight configuration when lowered to the ground. Stressing—stretching or compressing the pipe beyond its yield point—causes the metal to flow in a semi-plastic, ductile condition until it reaches its ultimate *tensile strength,* at which point it will rupture. The amount of plastic flow varies from one metal to another, and this is indicated in terms of *percentage of elongation.* Metals with a low percentage of elongation will rupture very quickly whereas those with a high percentage are capable of withstanding severe deformation such as a 1 1/2D bend.

Given the foregoing values for a particular material, we know what forces are required to bend it and how much bending it can withstand. Now we must verify the machine's capabilities, which are usually supplied by manufacturers' service manuals. Against the figures shown for the *machine bending torque* or section modulus, there is a simple formula for determining very quickly whether the machine is capable of bending the pipe. This formula is taken at 1D of radius, and is as follows:

Carbon steel: $8[OD^3 - (OD - 2t)^3]$ ton-inches
Copper: $5.5[OD^3 - (OD - 2t)^3]$ ton-inches
Stainless steel: $10.5[OD^3 - (OD^3 - 2t)^3]$ ton-inches

Additionally, these must also include the tooling factors which are:

Plug mandrel × 1.35
One-ball mandrel × 1.5
Two-ball mandrel × 1.6
Three-ball mandrel × 1.75

Applying this to, say, 4″ Schedule 40 carbon steel will give the following basic load limit:

$8 \times (4.5^3 - 4.026^3)$
$8\ (91.125 - 65.256)$
$8\ (25.869)$
206.952 ton-inches

For bends requiring a two-ball mandrel the final load will be $1.6 \times 206.952 = 331.12$ ton-inches $= 55{,}187.2$ **foot/pounds.**

How to Calculate Elongation

To calculate approximately how much elongation one will get in a particular piece of pipe for a given bend configuration, use the calculated difference between the developed *length* of the bend extrados and that of the centerline of the bend. For example, for a 1½D bend in 4½ OD material, it will be

$$= \frac{(OR \times 2 \times \pi)}{4} - \frac{(CLR \times 2 \times \pi)}{4}$$

$$= \frac{8.25 \times 2 \times \pi}{4} - \frac{6 \times 2 \times \pi}{4}$$

$$= 12.9591 - 9.4248$$

$$= 3.5343 = 35\% \text{ Elongation}$$

Wall Thinning

To determine the percentage of wall thinning expected for a particular bend, one can again use a simple formula. The extrados of the bend is stretched and thins out when a pipe is bent and the intrados is compressed and thickens, but the neutral axis or centerline remains virtually unchanged. If we take the outside radius of the bend, subtract the centerline radius, and divide the difference by the outside radius, we will get the percentage of thinning:

$$\frac{OR - CLR}{OR}$$

This in terms of 4″ pipe on a $1\frac{1}{2}$D bend would show something like this:

$$\frac{8.25 - 6}{8.25} = \frac{2.25}{8.25} = .27272 \text{ or } 27\% \text{ wall thinning}$$

The preceding example does not consider other factors, such as movement of a neutral axis from the geometric axis, pressure die assist or pressure die boost, tool drag, and lubrication, all of which can and will affect the actual amount of wall thinning achieved.

In most cases, however, benders can closely predict how much thinning to expect on a particular machine.

Springback

As mentioned earlier, the material along and on either side of the neutral axis or centerline is not stretched or compressed as much as the material in the extrados and intrados areas, which have been stretched and compressed respectively beyond their yield point. The centerline material, therefore, is trying to return to its original shape, yet it is constrained by the unyielding material on either side.

The effect of this is noticed immediately when the clamp and/or pressure die is retracted: the pipe tends to straighten out slightly, and this is what is referred to as springback. It must be considered from a bending viewpoint and compensated for in the machine setting, which must bend the pipe to the required angle of bend plus sufficient over-bend to compensate for or counteract the springback.

Springback is not a constant factor for all materials and the appropriate settings must be made for each type of material being bent. In fact, it is not uncommon to find that it may even vary slightly in the same material due to irregularities in the actual dimensions of the material itself. This is why many benders are only accurate to $\pm 1°$ for any bend angle. On opening the clamp or pressure die, since the material moves away from the die radius, the bent section is now somewhat larger. This is referred to as "radial growth" and is usually compensated for by a reduction in the actual centerline radius of the die.

Generally, for bends of less than 3D, there is no need to cut back the die to compensate for radial growth, but for material with high springback it may be necessary. It will be necessary as the radius increases, since the larger the radius, the greater amount of springback and radial growth there will be.

This again gets to be a serious problem for minimal-degree bends, and it becomes extremely difficult if not impossible to make bends of only a few degrees since the yield point is not attained, thus the pipe fails to reach the point where it will deform permanently.

From all this we can readily appreciate that it is simply not that easy to make a good bend accurately. It takes first-class equipment and first-class operators. Anyone can make a bend given the machinery to do so, but it takes a skilled, experienced operator, with good, precision tooling to make a really good bend. More will be discussed on the human element in another chapter.

True D Factors

Because pipe up to and including 12″ is always designated in nominal size, e.g., 6″, 8″, 10″, and yet has a larger than nominal true outside diameter, the so-called 1½D radius in all the IPS sizes is misleading since the D factor, from a bending and tooling viewpoint, is always taken as R/OD. Thus, the true D is something less than the nominal. This is shown clearly in Table 1-1.

In fact, in the 1½D, 2D, 3D, and 5D configurations for IPS pipe we see some quite interesting differences and these differ-

Table 1-1
D Factor

Nom. Pipe Size	True OD	1½D Radius	True D Factor	True 1½OD Radius
1″	1.315″	1.5″	1.14	1.9725″
1¼″	1.66″	1.875″	1.129	2.49″
1½″	1.90″	2.25″	1.184	2.85″
2″	2.375″	3.00″	1.263	3.5625″
2½″	2.875″	3.75″	1.304	4.3125″
3″	3.5″	4.5″	1.285	5.25″
3½″	4.0″	5.25″	1.3125	6.00″
4″	4.5″	6.00″	1.333	6.75″
4½″	5.0″	6.75″	1.35	7.50″
5″	5.563″	7.5″	1.348	8.344″
6″	6.625″	9.00″	1.358	9.9375″
8″	8.625″	12.00″	1.391	12.9375″
10″	10.75″	15.00″	1.395	16.125″
12″	12.75″	18.00″	1.411	19.125″
14″	14.00″	21.00″	1.5	21.00″

ences for the thinner wall schedules can have a bearing on the tooling selection, possibly the difference between a three-ball or two-ball mandrel and the need for a wiper die as illustrated earlier.

Thus, when selecting the tooling for a particular bend, the proper factors must always be used. These factors in reference to the tooling charts will usually provide the correct answer.

Figure 1-2 shows some examples of so-called "standard" bends, all of which are on a single plane, although those with more than one bend direction are "offset plane" bends. Where second bends lie in a plane other than the plane of the first bend,

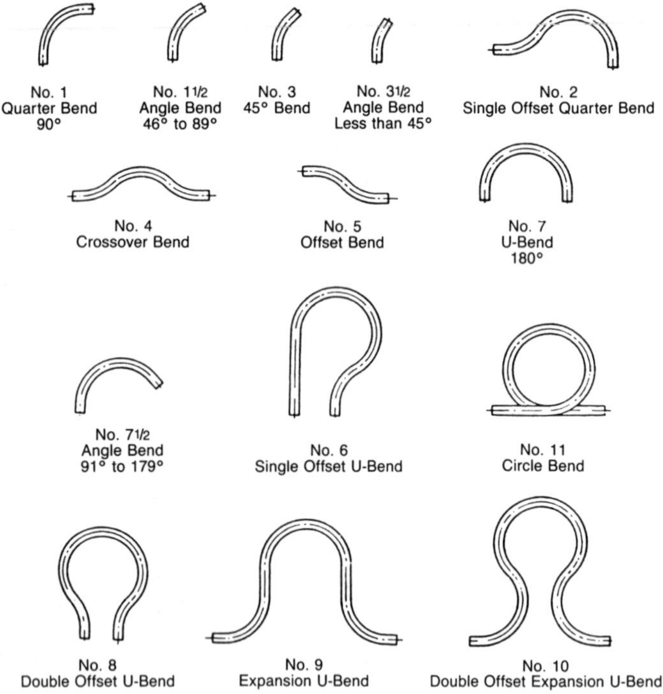

Figure 1-2. Examples of "standard" bends.

they are referred to as "multiple plane bends." Where the first tangent point of the second bend *starts* at the rear tangent point of the *preceding* bend, without any straight in between, or a sufficient amount of straight for normal clamping, they are called "compound bends." Even though the tooling might not be available to do these bends as a single workpiece, the configuration shown can still be made by making the bends separately, trimming off the tangents, beveling the ends, and then welding the bends together. This, of course, is a more time-consuming and costly method of fabrication than having the machinery and tooling to make the configuration in a single length of material.

2

Types of Bending

Compression Bending

Compression bending is a process whereby pipe or tube is bent to reasonably tight radii, usually without the use of mandrel or precision tooling.

It is accomplished by clamping the workpiece behind the rear tangent point and then by means of a rotary arm rolling or compressing the material around and onto a die. This effectively puts the material under radial compression and provides for tighter bends than would be possible by press bending. Another method is to use a follower type die instead of a roller, but the principle is the same.

If a tight bending radius is not required and the geometry of the bend shape is not critical, the compression bending method is the simplest and most economical process. In contrast to radial draw bending, the clamp and die remain stationary. Bending is accomplished by wiping the workpiece around the die.

The upper illustration in Figure 2-1 shows the roller principle and the lower illustration shows the follow block principle. Re-

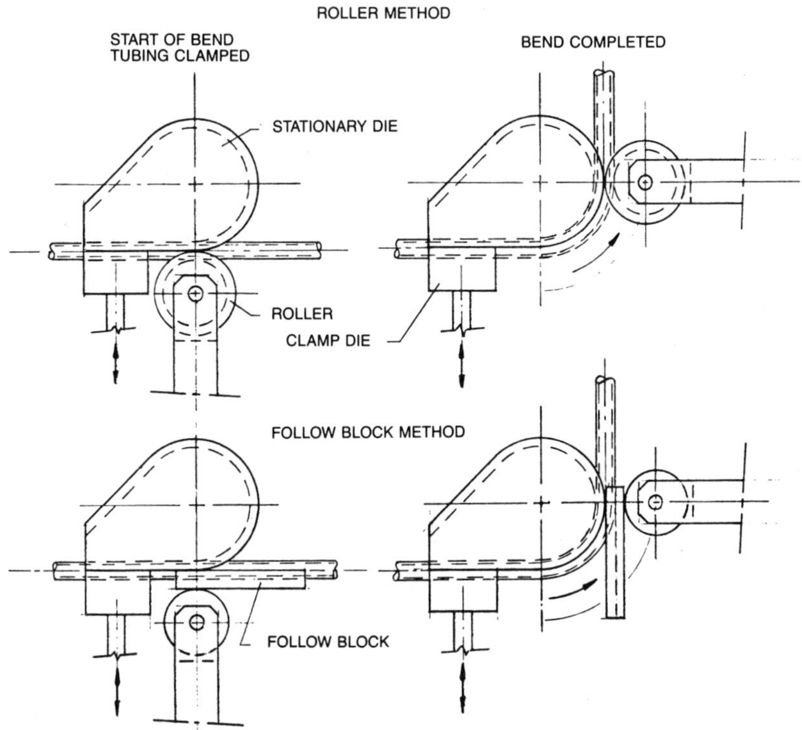

Figure 2-1. Schematic showing compression bending tooling and bending action.

gardless of which method is used, it is important that the forming action start as close as possible to the tangent point. When bending round or square tube to close radii, the collapse or crush bend methods are used.

Because the ovality and wall thinning in compression bending is somewhat greater than in draw bending, the process is used mainly for noncritical application. It is particularly suitable for medium to heavy wall material.

Wrinkling on the inside of the bend, necking, and excessive ovality are the more common problems in using this method, particularly as the radius gets smaller. For generous radii this is still a quick method, however, and in heat exchanger tube bending it is the one most commonly used. This will be dealt with further in Chapter 3.

One advantage of a compression bender with a rotary drive—as opposed to a converted draw bender with a single direction of bend drive—is the ability to switch the bend direction by replacing the clamp arm on the opposite side and reversing the roller arm bend direction. Some multi-purpose machines available today can be used as draw benders or compression benders, or even as coil benders simply by changing certain components. For job shop work, these machines are ideal as they can perform all three functions rather than the user having to have three separate machines.

Ram Bending

Ram Bending is perhaps one of the oldest and simplest methods of mechanical bending. The principle is akin to bending a stick across one's knee.

The principle of ram bending is illustrated at the top of Figure 2-2. Ram action forces the pipe against two fixed rollers or pivot blocks to bend around the die. Large sweeping curves can be bent in small increments, moving the pipe for each bend.

Although there are considerable limitations to using this method in terms of minimum radii achievable, it does have the advantage of being able to bend quite large-diameter pipe at relatively low cost, particularly for bends with radii 5 or 6D and larger.

Ram benders are used extensively in the field for smaller-diameter pipe bending, since they are readily portable and easy to set up and operate (Figure 2-3 and 2-4). The principle is simple: The pipe is positioned in a fixture or jig with the ends resting against rollers or pivot blocks. A hydraulic ram or other method of actuation then pushes against the center of the pipe—which lies in a die at the front end of the ram—until the required angle of bend has been achieved. The fixture has predetermined holes for the rollers or swivel blocks which are used with the appropri-

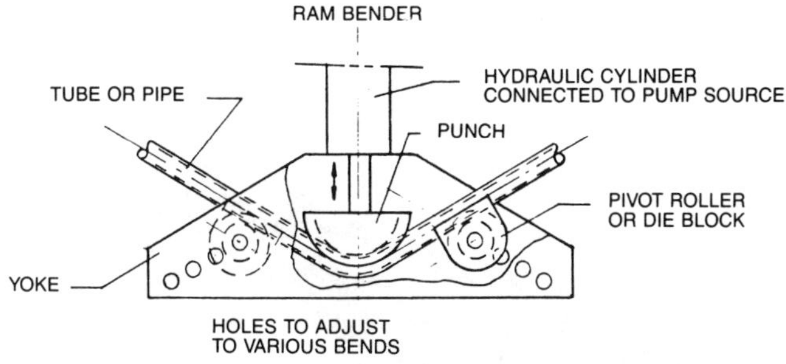

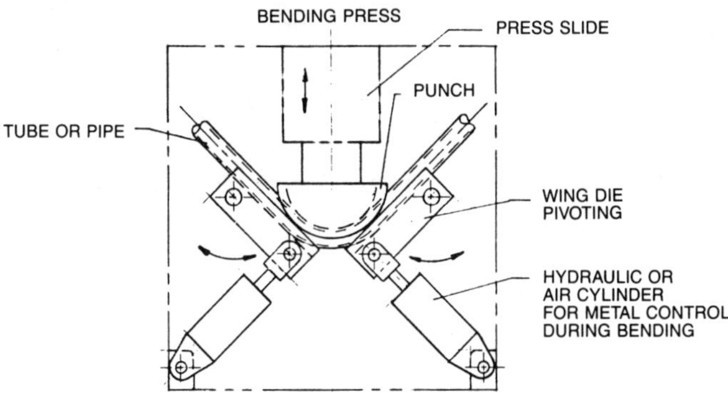

Figure 2-2. Schematic showing press bending tooling and bending action.

ate radius die or shoe. Radius and pipe size changes can be effected very quickly and simply. A number of manufacturers supply this type of bending equipment in sets, complete with all the necessary dies, rollers, and accessories.

The larger versions are usually permanently located in a pipe shop since they are too large and heavy for field use (Figure 2-5). Some of these can accommodate pipe up to 20″ in diameter with wall thicknesses of 1/2″ or more.

Press bending, employing the fully supporting wing-type tooling as shown in the lower illustration of Figure 2-2, is a combination of ram and compression bending. High-production hydraulic

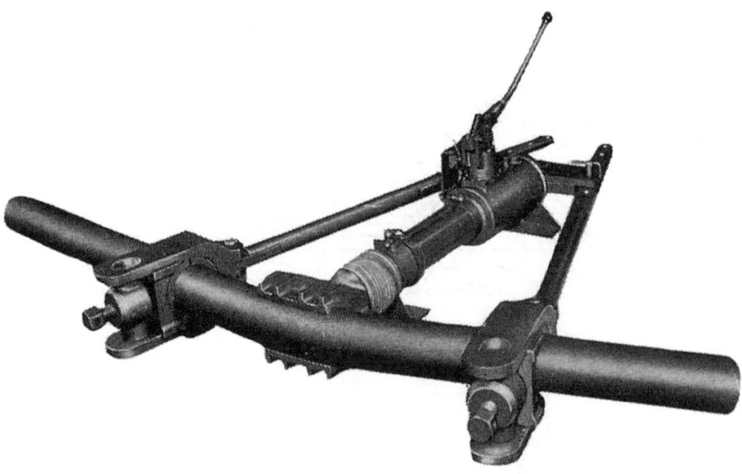

Figure 2-3. Typical field or shop ram bender for large pipe—manually operated.

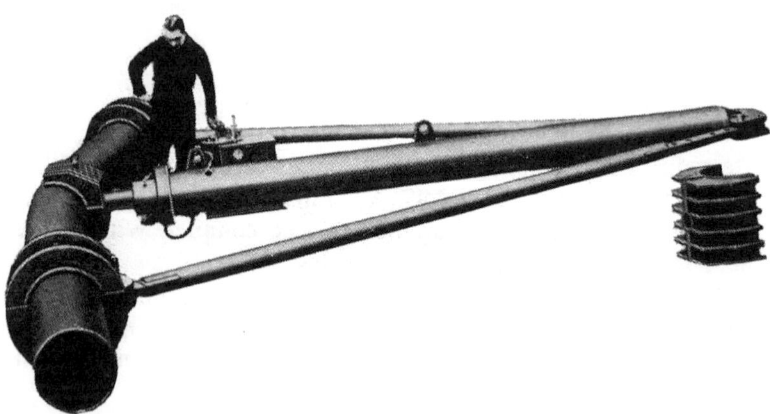

Figure 2-4. Field ram bender utilizing an electric motor drive for the hydraulic ram.

Figure 2-5. Shop ram bender. Note the "shoes" at each end. These pivot during the bend while the ram pushes the shoe in the center of the pipe. (Courtesy of Wallace Manufacturing Co.)

presses are used for high-volume items, such as automotive exhaust pipes, bicycle components, and metal furniture parts. Since bend angle is controlled by the amount of die descent, such presses are easily programmable. Bends are made three or four times faster than on other types of equipment. Press bending is limited to bends of less than 165° bend angle; also, the tube is reduced slightly on the inside of the bends. Such presses are ideally suited for "crush" or "collapse"-type tube bends.

For noncritical work, ram benders are ideal because of their simplicity. One can find this type of equipment in use throughout the world, some units dating back before World War II. But as design and engineering demands increase, with insistence on more precise tolerances and quality, one must turn to more sophisticated machinery. Nevertheless, for field application the ram bender will likely be around for a long time to come.

Draw Bending

This is the most widely used method of bending pipe and tube today, particularly for tight radii and for thin wall material. Advantages include maximum control of wall thinning and ovality. Draw bending usually involves the use of mandrel tooling, but in certain instances when bending heavy wall tube to tight radii, use of the mandrel may be omitted. In fact, empty bending (as it is referred to) is achievable down to radii as tight as 1D under certain conditions.

The principle is the same regardless of the make of machine. The pipe is clamped just beyond the front tangent point against a die. Then, usually by means of a swing arm, the material is drawn over a mandrel inside the pipe. At the same time as it is being drawn around the die, a pressure die, either static, rolling, or boosted, transfers the material to the die at the tangent point.

One of the main advantages of the radial draw bending method is that the material is being "drawn" into the bending area past the tangent point. A suitable internal mandrel can therefore be placed at the tangent point to prevent collapse of the tube or pipe during the bending action.

The upper illustration in Figure 2-6 shows the typical bending action when using a static pressure die. Since material must overcome the combined friction of wiper die, pressure die, and internal mandrel when being drawn into the bend, the static pressure die feature is not ideal when bending thin wall tubing to close radii.

The bottom illustration in Figure 2-6 shows essentially the same bending action using the "following pressure die" principle. The pressure die slides on rollers at the same rate as the pipe is being drawn into the bend. A further improvement to this principle, when bending thin wall tubing to close radii with minimum wall thinning, is the addition of a hydraulic boost cylinder for the pressure die in the direction of bend. This results in the material being pushed and drawn into the bend during the bending cycle.

The mandrel, which may be either shaped or plug or with one or more balls, assists in forming the bend while controlling ovality by forcing the material in the bend extrados over the nose of

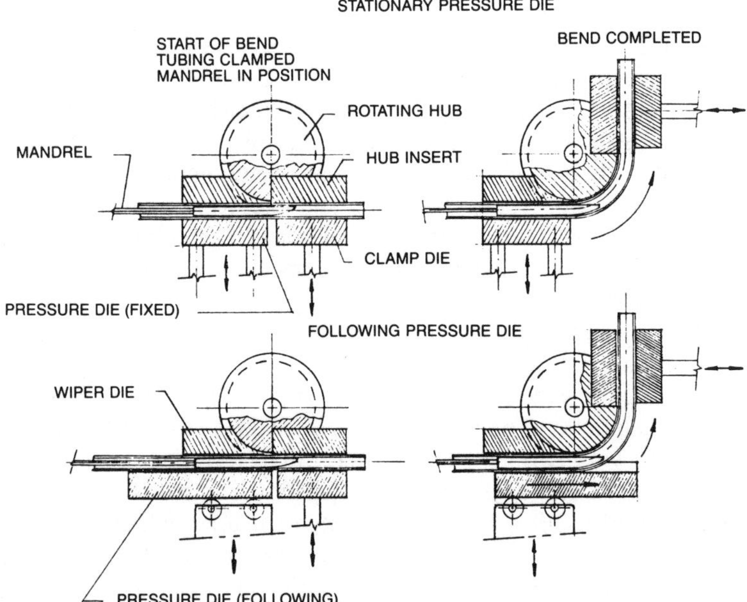

Figure 2-6. Radial draw bending—tooling and bending action (schematic).

the mandrel. The balls are designed to maintain the ovality after the material passes over the nose of the mandrel. The balls are never—or shouldn't be—made to function as the mandrel itself. To maintain ovality throughout the arc of the bend in ultra thin wall tube on critical radii, there may be as many balls as needed for the full developed length of the bend.

The mandrel itself is a solid piece, but the balls are articulated either on a single plane or with a 360° directional capability. The mandrel is attached to a mandrel rod which runs the full length of the machine and is itself attached to a cylinder which can move the mandrel forward and backward.

On completion of a bend, the mandrel is retracted before unclamping and removing the workpiece. Then it is moved forward again before the next bend is started. The position of the mandrel nose in relation to the tangent line is absolutely critical and has a very definite effect on the bend quality. Generally, the nose projects just past the tangent point, since this is where the bend starts and keeps the outer wall from moving inward toward the die.

Most draw bending machines today are equipped with a mandrel lubrication system whereby suitable forming lubricant is pumped from a reservoir through the mandrel rod into and out of the mandrel itself. This lubricates the inside of the tube as it passes over the mandrel before reaching the tangent point and the area of greatest pressure where the mandrel is forcing the extrados to maintain its ovality.

For critical bends, particularly in thin wall material, a wiper die is usually also necessary. This is shaped rather like a solid shoehorn with two very critical radii involved. One matches the radius of the die itself, and the other matches that of the OD of the material. Both radii meet at the tip or feather, which on a new wiper is razor sharp and must always be protected. In use, the feather gets eroded, particularly when bending coarse materials, but the mandrel can be reworked as required, which is less costly than buying a new one each time the one in use gets worn.

Thus we have five basic items in draw bending tooling: the die or center former, which also has either an integral or removable clamp section; the clamp; the pressure die, sometimes referred to as the follower die; the mandrel; and the wiper die.

Fully tooled rotary draw bending includes all these tooling components. See Figure 2-7.

The clamp supports the pipe against the straight section of the bend die. Clamping surfaces usually are knurled or sand blasted to provide maximum grip. The pressure die supports the straight portion of material while it is being "drawn" into the bending area. Depending on the type of machine, the pressure die is either static or it moves at the same rate as the material is drawn into the bend. The wiper die guides and supports the pipe to a feather edge close to the tangent point of the bend die. The main function of the mandrel is to prevent excessive collapse of tube or shape during the bending cycle. A plug is used in some cases to prevent distortion of the tube end in the clamp area.

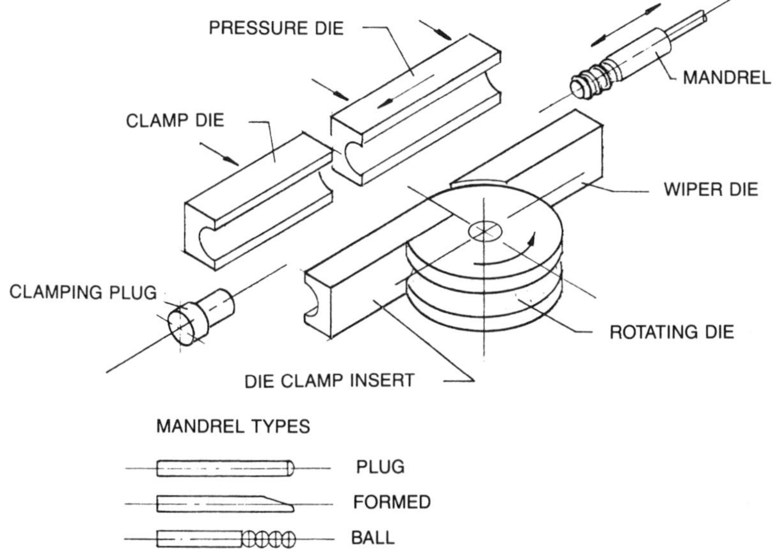

Figure 2-7. Radial draw bending—tooling features.

All tooling components for high-volume production bending should preferably be made of hardened tool steel. Tooling which has sliding contact, such as pressure die, wiper die, and mandrel, should, in addition, be ground and polished to a fine finish.

Incidentally, the wiper is called a wiper *not* because it is designed to wipe out wrinkles, but because its purpose is to prevent wrinkles from forming, and because it wipes against the inside of the die itself.

One more item of tooling, which only applies to draw benders with a boosted pressure die is a boost clamp. There are several varieties depending upon the manufacturer, but again, the principle is common.

Boosted means the pressure die has a cylinder or some other mechanical means of pushing it forward as it travels with the material as a means toward controlling wall thinning. If the boost pressure exceeds the rate of travel of the tube or pipe, the pressure die may slip. To prevent this and to obtain greater boost forces, the tube is clamped firmly against the pressure die itself. Thus the tube can be pushed harder into the bend area.

In the hands of a skilled operator this can be a very effective method of controlling wall thinning. But the problem of opposing forces is critical and a careless operator can ruin a bend very easily by boosting the pressure die too much and either forcing the tube through the clamp section in the early stages of the bend or, worse still, buckling the tube and possibly even damaging the tooling and/or the machine.

A boosted pressure die should never be confused with an assisted pressure die. The latter has forward movement assistance by means of a cylinder. It is classified as boosted only when the bend material is clamped to the pressure die and considerable force is induced to move the material forward during the bending process.

In addition to the tooling, the best draw bending machines today have a number of other characteristics—such as a carriage—to assist in loading the tube by drawing it back over the mandrel and, after completing a bend, moving the tube forward to position it correctly for the start of the next one. Most draw benders equipped with a carriage also usually have the means to rotate the tube for a change of plane in either direction. Thus, a machine with this type of equipment would have a three-axis capability—DOB for the required angle of bend, DBB and POB. It can be operated manually for settings in all three cases or (on CNC machines) be programmed for the settings automatically. Further refinements such as a swing-away wiper die, wiper lubrication system, flanged pipe chucking, and other such items are all available to improve bending capabilities of a particular machine.

But when all is said and done, it still comes down to the operator. If he lacks the necessary knowledge and skills, one can have the finest machine in the world and still produce poor-quality bends.

Each manufacturer usually develops one or two special features on its machines that sets them apart from the competition, at least for a while, until the competition comes out with a similar item.

One U.S. manufacturer, Coast Iron Works, has a hydraulic clamping system that is mounted directly onto the die, thereby eliminating the need for a swing arm. It also helps keep cost down. The clamping unit is removable and interchangeable for use with different size dies and is equipped with inserts for pipe size reduction on the same radius. This system has several obvious advantages, but for any compound bend requirements, i.e., starting a second bend at the rear tangent point of the preceding one, it does present a problem.

The unique overhead hydraulic clamping system in Figure 2-8 is a feature on many Coast Iron Works' machines and effec-

Figure 2-8. Clamp and boost system. (Courtesy of Coast Iron Works, Inc.)

tively eliminates the need for a swingarm. Note also the overhead clamping system on the pressure die for booster bending.

One feature that a German manufacturer has on all its machines is straight-through chucking. This provides the means to rear feed a length of material longer than the machine itself if empty bending is to be done. (Unfortunately, this does not seem to be the case on U.S.-made draw benders.) The mandrel cylinder is moved to one side and the material, fed from the rear, is moved forward by the carriage. This is called *hitch feeding.* For empty bending 40′ or longer boiler tubes this procedure saves having to bend the rear 20′ section, remove the tube, reload it, and bend the front section. An added advantage is that with a flying cutoff on the bender one can use longer—even continuous—lengths of pipe or tube and simply cut after bending. Handling is always a problem. Not too bad for small diameter materials such as 2″ Schedule 40 pipe, but handling, loading, and unloading after bending a 20′ length of 6″ Schedule 40 or 8″ Schedule 40 pipe can be a real headache—380 lbs of pipe for the former and 570 lbs for the latter. This means using a crane or a lot of plain ordinary muscle power.

In situations such as this, an autoload system is well worth the investment. Apart from being much faster and safer, it is far less labor intensive. See Figure 2-9.

Getting the bend material off the machine is also a problem. Not too bad for a single bend in a 20′ joint but for several bends with changes of plane it is a tedious task. Robotics and computers will resolve this problem in due course. In fact, this approach has already been taken by one company which uses two robot arms—one to load the material, the other to remove it—and the computer calculates out the point of balance in the bent workpiece so that the robot arm clamps the pipe at just the right spot. See Figure 2-10.

Draw bending machinery today is highly sophisticated and becoming more so all the time. It is a far cry from the simple draw bender of only a few years back with manual controls for bend/bend return, pressure die forward/return and mandrel forward/return. Even the clamp was a manual operation.

Figure 2-9. Feed table and auto-load system in conjunction with a flanged-pipe bending machine. (Courtesy of Schwarze-Wirtz A.G.)

Figure 2-10. Robots used in tube bending. The upper robot arm loads the straight tube while the lower one removes the finished bent workpiece. (Courtesy of Schwarze-Wirtz A.G.)

Figure 2-11 shows a modern, high-speed CNC Bending Machine. In this picture the pipe has been rotated 180° in preparation for the next (offset plane) bend, while the drop-away clamp has retracted for return of the swingarm as the material is moved forward into position for the next bend.

The control panel for a modern automatic bender is quite complex and for a CNC machine, even more so, particularly with such added refinements as self diagnostic capability. Selection of the right machine for one's particular requirements can be a complicated, lengthy process.

The machine in Figure 2-12 has up to ten bend settings with automatic movement forward (and backward) of the material for DBB settings. Note also the dropaway clamp.

Some manufacturers are now differentiating between a draw bender with boosted pressure die and a booster bender, which lit-

Figure 2-11. CNC draw bender. (Courtesy of Eaton-Leonard Corp.)

erally pushes the pipe forward without it necessarily being clamped directly onto the pressure die. Since the pipe is not drawn, but rather is pushed into the die, the manufacturers claim this type of machinery should be classified as a booster rather than a draw bender. Technically they are right.

The difference between the two should be predicated upon the amount of material needed to make a bend. Since draw bending stretches the material and requires *less* than the calculated developed length of the bend, booster bending, like induction bending, compresses the material and should require *more* than the calculated length.

Nearly all bending machine manufacturers are constantly seeking ways to improve their products, reduce production time and cost and enhance finished product quality. Space does not permit me to list and illustrate all the many improvements that have

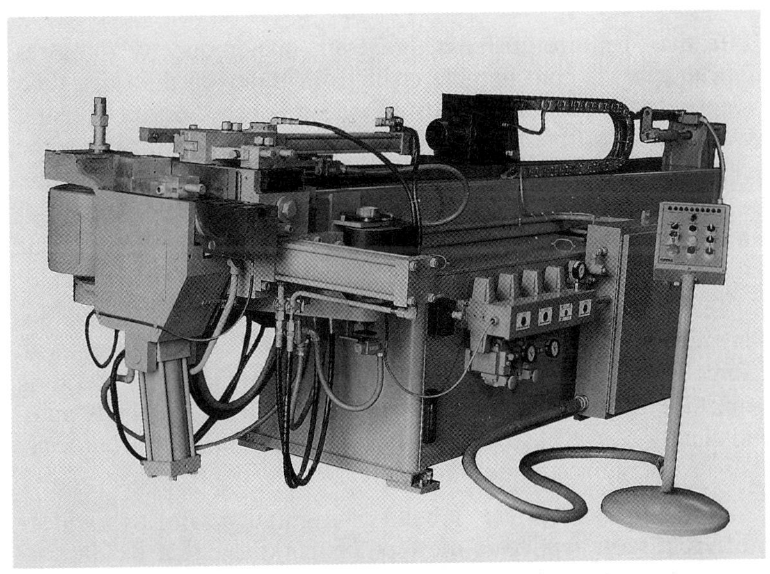

Figure 2-12. 10-bend draw bender. (Courtesy of Conrac Corp.)

taken place during the past years. Each one has contributed significantly to technological advancement either generally throughout the industry or in certain specific areas.

Perhaps the most recent improvement is the introduction of a multiposition bend head for multiradius application. Although this was conceived essentially for tubular furniture bending purposes, it will no doubt have equal application in the boiler tube field where multiradius configurations are common in a single length of tube.

Tiered tooling has been in use extensively for a good many years, but has generally been confined to relatively close changes of radius and/or material size.

Figure 2-13 illustrates the latest advance in this area. There are three bend configurations shown; the middle one is for straight clamping while the top and bottom sections are for compound bends—the upper die shows a previous single bend clamp section and the lower die shows a double (offset) bend clamp section. This provides for a three bend configuration with a single tooling set up.

In the furniture business, however, one frequently finds requirements for considerable changes of radius in the same tube length. In the past this was usually accomplished either by using two entirely different machines, one for each radius, or in some cases a complete change of tooling set up on the same machine.

Both methods were equally effective insofar as the end result was concerned, but both required too much time and effort to be really cost effective.

Eaton-Leonard Corporation, renowned for its innovative approach to bending, addressed this problem and came up with some remarkable developments that now enable a CNC bending machine to produce compound and/or multiradius bends automatically as well as bends in two planes in not only round but also rectangular and oval tubing as well.

Figure 2-14 illustrates a bend configuration with two widely different radii applied to the type of workpiece that would normally have needed either two separate tooling changes or two different machines.

Figure 2-13. Tiered, compound-bend tooling. (Courtesy of Eaton-Leonard Corp.)

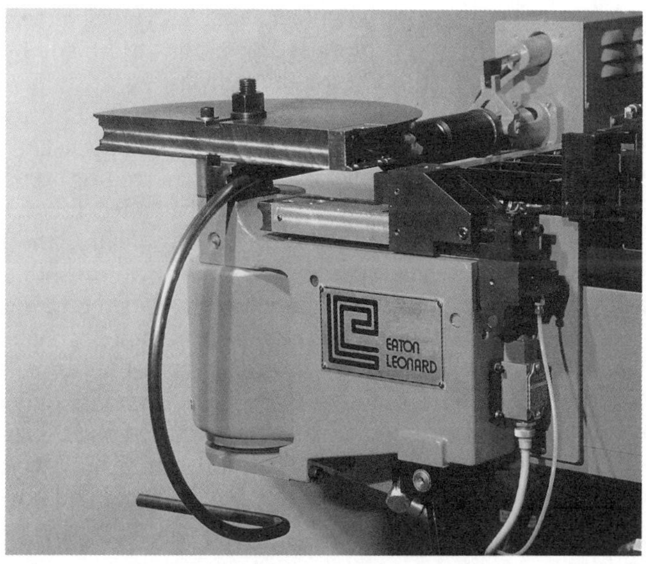

Figure 2-14. Tiered multi-radius tooling. (Courtesy Eaton-Leonard Corp.)

For round or square tubing, multiple bend dies and associated tooling are not too much of a problem, but for bending rectangular or oval tubing in multiple planes, different pressure dies are required as well as, under certain conditions, wiper dies and mandrels.

Hitherto on manually operated machinery this type of bending, while equally effective in terms of end results, would have taken a great deal of time and labor input to achieve. Now, however, with CNC machinery it can all be done by one machine on a single tooling set-up, supervised by one man in a fraction of the time it used to take.

Coiling

Making coils is an art unto itself. Tight radius coils are often made on draw bending equipment or, for small diameter pipe and tube, by winding the material onto a suitable cyinder to form the coil on a drum. However, in most cases, coils are made by machinery designed specifically for this purpose.

The procedure for coiling is relatively simple. The pipe is fed between three rollers—the center one of which is usually the drive roller while the other two are either powered or are idlers. On some machines a clamping roller may be added for tight radii. This fourth roller holds the pipe directly against the drive roller in order to reduce slippage. On others one or both idlers may be driven. The position of the idlers dictates the radius and the more they are advanced the tighter the radius becomes.

Two types of roll benders are in general use: the "three-roll" (pyramid type) which is built in both vertical and horizontal versions. Some types are universal and can be placed both vertically and horizontally to suit a particular job situation. The "two-roll" (pinch type) bender is likewise built in both vertical and horizontal versions. The bending action of both types can be seen in Figure 2-15.

This procedure results in a horizontal plane of bend only and is excellent for large and very large radii beyond 90° or 180° (Figure 2-16). However, sooner or later the front of the pipe will come around toward the back end to make a full circle. Obviously, this cannot be done if there is still material feeding into the machine so one either stops or, for large radii, lifts the front end to pass on top of the rollers and continue on around again. This creates a continuous coil on a one-diameter pitch, i.e., the coils

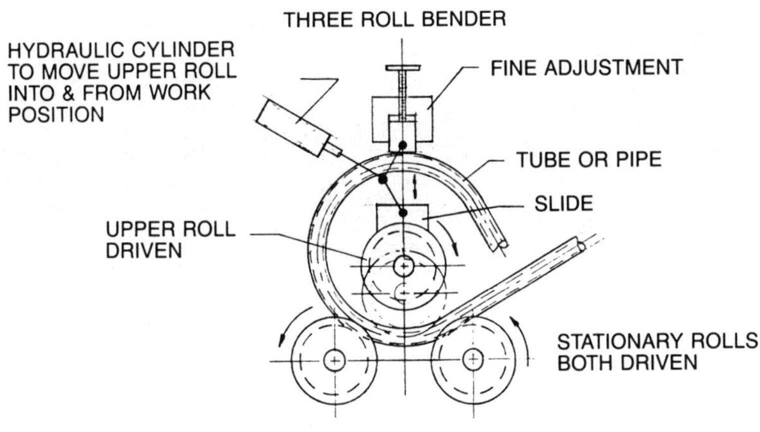

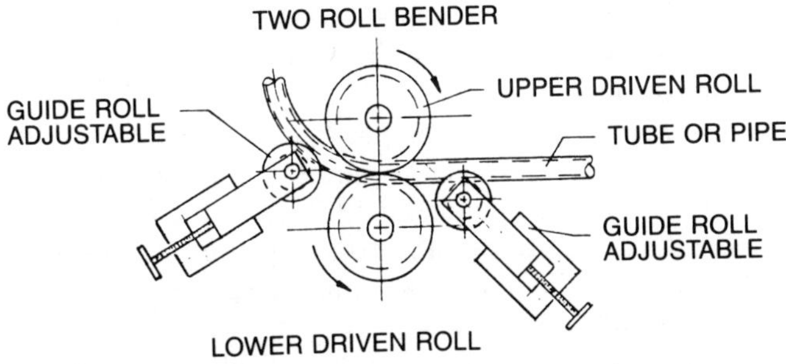

Figure 2-15. Roll bending—tooling and bending action (schematic).

Figure 2-16. Roll/coil bender. (Courtesy Wallace Manufacturing Co.)

touch all the way around and the natural spring of the material keeps them closed tightly together. Should it be necessary to have a greater than one-diameter pitch, then an additional roller must be used, which forces the pipe upward as it comes out of the second idler. Adjustment of this roller will create the desired amount of pitch.

For critical diameters and particularly for material with a high springback factor, it may be necessary to add one more set of rollers—also idlers—at about 180° from the tangent point of the drive roll, to insure that the coil stays on the required radius.

As with every bending machine there are limitations. Coiling is an 'empty bending' process in most cases, therefore, it is not possible to form extremely tight coils when the bend would normally require tooling in order to maintain ovality tolerances.

For situations such as this, one can use a draw bender with mandrel tooling and special clamps made to the radius of the coil. Turns are usually made 90° at a time and pitched accordingly.

The simplest way to make a coil, particularly in small diameter pipe or tube is to wind it into a drum (Figure 2-17). The drum OD is generally less than the required coil ID in order to allow for springback. The beauty of this type of coiling is that if the

Figure 2-17. Drum rolling. Note the crude guide roller. With this type of coiling the drum diameter can be increased by winding on a suitable thickness of strap material if the drum of the required OD is not readily available. (Courtesy of Tubebend Ltd.)

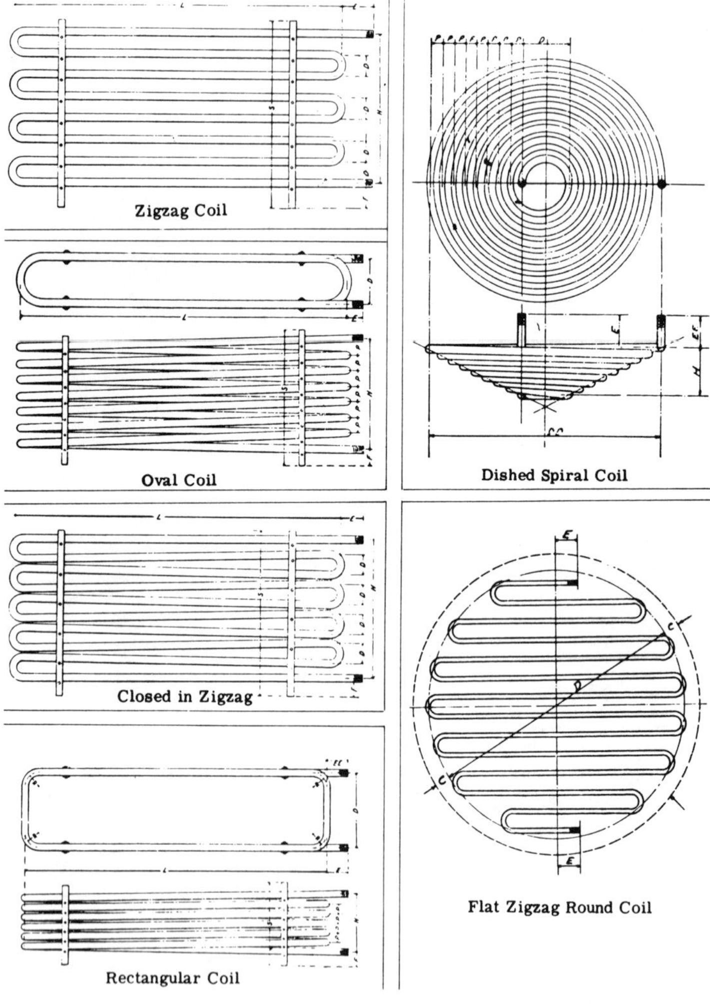

Figure 2-18. Typical coil configurations.

diameter needs to be increased one can always wind on some thin strap, 1/8″ or 3/16″ or 1/4″—either a single layer or two to three layers as needed—to increase the OD of the drum.

Pitch is controlled by a single roller on a vertical slide and as the drum rotates, so the material is fed onto it at the required pitch distance.

In many cases, coil configurations require specially designed machinery, particularly where quantity production or frequent repetition is involved. The coil configurations in Figure 2-18 illustrate only some of the many varieties possible.

Internal Roll

Perhaps the least familiar bending machinery in common usage today is the Rotoform™. This is a patented French design machine manufactured, tooled and serviced under license by Pines Engineering Co. in the United States. See Figure 2-19.

There are not too many of these machines around, perhaps because although they have certain advantages over the more conventional types of machinery, they do have some drawbacks.

The process itself is a departure from the norm and uses the AREF (Automatic Roll Extrusion Forming) principle whereby controlled pressure is applied to the inside diameter of the pipe by means of a rolling head rotated within the pipe itself.

The main advantage of this system is that it can produce almost any desired radius within the minimum/maximum bend parameters of the machine. Thus, changes of radius can be accomplished on the same piece of pipe without the need for a tooling change. Compound bends, and bends in excess of 180°, can be made as well.

One of the principle disadvantages is that radii of less than 3D are not achievable and the pipe sizes are limited to the 5–12″ IPS range. Another disadvantage is the greater than normal (in contrast to draw bending) work hardening and wall thinning of the material, particularly in the extrados of the bend. On the other

Figure 2-19. Internal roll bender. (Courtesy of DeKay Fabricators, Inc.)

hand, the ability to handle thin wall (Schedule 5) pipe in all the sizes accommodated by the machine is a distinct advantage, particularly in the upper size range of 8″-12″ IPS.

For air conveying systems in thin wall aluminum or stainless material where radii are in the 7-10D range and line pressures are low—12-15 PSI—the means for producing almost any desired radius outside of the so called "standards" is a distinct advantage and can save thousands of dollars in tooling costs.

For example, a common radius for 8″ air conveying pipe is 72″ (9D) but if, for one reason or another, bends are needed on a radius of say 62″ and a die for this particular requirement is not available, the cost of making one—or having one made—is prohibitive and can easily run up to thirty or forty thousand dollars. Plus, of course, there is the loss of time while the die is being made. Unless a very large number of bends are required for the "odd-ball" 62″ radius so that the die cost can be effectively amortized, it is not worth the time and effort, let alone the cost. It is characteristic of "odd-ball" radii that they are specified for a minimal number of bends only.

The Rotoform™ bender covers an area of pipe bending that is not fully covered by either cold or induction bending and has a place in the industry. Cold bending machines over 8″ in size are few and far between, and even these seldom go beyond 5D in radius capability. Induction benders, particularly the new generation of smaller ones, cannot accommodate thin wall (Schedule 5 and Schedule 10 pipe) much below about 5D and, at the present time, cannot handle non-ferrous material such as aluminum or copper.

Thus, the Rotoform™ process fills a gap left by the other processes. For a pipe shop that concentrates only on bending up to 12″ IPS, a good combinaton of machinery might be as follows:

One 4″ radial draw bender for 1½D to 3D
One 8″ radial draw bender for 2D to 5D
One Rotoform™ bender for 3D to 10D
One Induction bender for 2″ to 12″ IPS
One 4″ or 6″ coiling machine

These five units would then cover virtually the complete spectrum for bending of all sizes and all types of materials from 1″ to 12″ IPS.

Several benders/fabricators in the United States have some of these units in their shops, but none have all of them and any company that did would certainly have a unique range of capability.

Induction Bending

The induction bending process is a relatively new technology that originated in Czechoslovakia. In 1967, certain rights to this process were acquired by Verkamaf Holland BV, a Dutch engineering group.

Induction bending first appeared in the United States in the early 1970's, when a 60″ machine was constructed and put into operation by Pipe Benders, Inc. of Duluth, Minnesota for their own in house use, followed about a year later by a 28″ machine. The first commercial machine was a PB 600 manufactured by Cojafex BV, a member of the Verkamaf Group. It had a capacity range of 6″–28″, and was installed at Tulsa Tube Bending Co., Tulsa, Oklahoma in 1977. Three years later Benjamin F. Shaw, Laurens, South Carolina, and Associated Piping and Engineering (now Johnson Controls), Clearfield, Utah each installed a PB 850 model for bending pipe in the 8″–32″ size range, and the following year North American Specialty Pipe, Calgary, Alberta installed a 36″ machine. At about the same time, Associated Piping and Engineering added a second unit. This was a PB 1600 for bending pipe up to 62″ in diameter and is believed to be the largest machine of its kind in the country. In Europe, however, where there are more than 20 Cojafex machines in operation, Fabricom, a Belgian Company, installed a monstrous 70″ machine in 1982. This was their fourth acquisition.

All these machines were designed and manufactured by Cojafex, which has more than 30 machines in 16 different countries and 5 continents of the world.

Cojafex, however, is not the only manufacturer. There are three others as well, two English companies and a Japanese one, manufacturing induction benders. The latter, Dai-Ichi High Frequency Co. Ltd., has machines in Brazil, Russia, China, and Korea as well as a dozen or more in Japan itself.

In England, two companies—Gregson Pipe Work Ltd. in South Shields, and Pipework Engineering Development in Tipton—manufacture induction bending machines. The former also has machines in Scandinavia as well as one in the United States.

Until 1982, virtually all these machines were designed for large diameter pipes. Very little consideration was given to producing machines for bending smaller pipe diameters, largely because this area—1½″–8″—was already covered by the cold bending process. Gregson and Dai-Ichi had both produced smaller machines in the 1″–12″ range, primarily for their own in house requirements, but none of the four manufacturers had designed and built a smaller machine specifically for foreign markets.

Then, in 1982, Gregson and Cojafex each delivered a small machine to the United States covering the 2″–12″ size range. The Gregson machine went to Houston Pipe Benders and the Cojafex unit went to a newly formed company, International Piping Systems Ltd. in Baton Rouge, Louisiana where a PB 850 was already on order.

The Cojafex unit introduced some remarkable new features, chief of which was the ability to make 1 1/2D bends with wall thinning controlled to within the tolerances of a butt weld fitting. This represented a real technological breakthrough. For years the cold bending machine manufacturers had been trying to crack the barrier of making a 1 1/2D bend within the tolerances of a fitting. While they had, for the most part, been able to do so in certain materials such as A-587 carbon steel pipe, they could not manage to get down to 12 1/2% or less wall thinning in A-106 and A-53 on a consistent basis.

There is no doubt that this achievement on the part of Cojafex will herald significant changes in piping design and fabrication since the 1 1/2D bend now becomes, in essence, a weldless fitting.

Technology of Induction Bending

The basic principle of induction bending is a radical departure from both the hot slab method and the cold draw process. For the former, the entire sand-filled bend area is heated. For the latter, the pipe is slightly stretched while being drawn over a mandrel. Induction bending, on the other hand, compresses the pipe while passing it through a narrow heat band.

The pipe is contained in the body of the machine and clamped at the front end, with a pusher bar or clamp at the back end. The clamp is mounted on a bending arm which pivots up to 180° from a fixed point. The distance between the pivot point and the center of the clamp determines the radius of the bend, and is variable to within 1/10″. See Figure 2-20. On the large Cojafex machines, the main frame is mounted on transverse rails and self

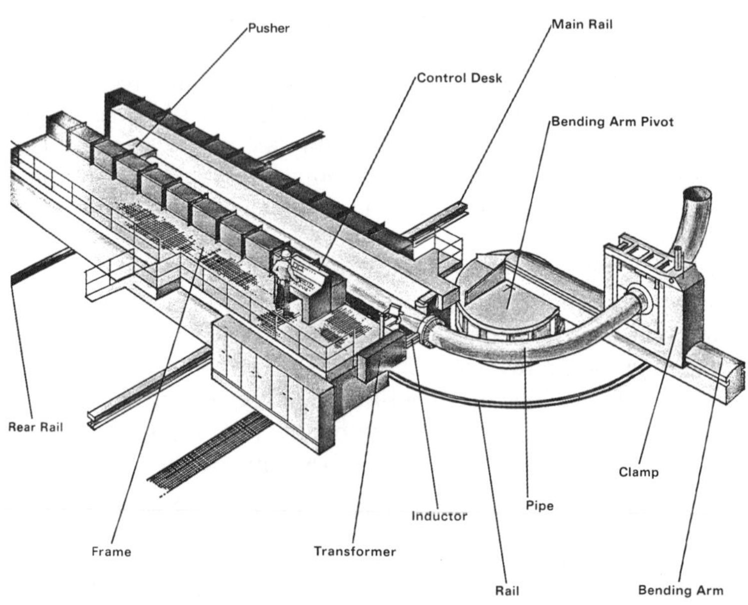

Figure 2-20. Large-diameter induction bender. (Courtesy of Inkamaf Corp.)

adjusts to the radius setting of the swingarm. On the small Cojafex machines, as on Gregson and other makes, the pivot point is adjusted either by travel transversely on the main frame or by a carriage extending away from the main frame at right angles to it. See Figure 2-21.

When everything is aligned with the pipe properly secured in its clamps, electric power is transmitted to an inductor ring, which circumscribes the pipe at the tangent point. This causes the pipe to heat up in a relatively narrow band immediately beneath the ring. This heat band is controlled, to some extent, by water cooling or quenching on the far (clamp) side. See Figure 2-22.

The temperature of the heat band is monitored by two pyrometers, which transmit the information to the machine controls. As

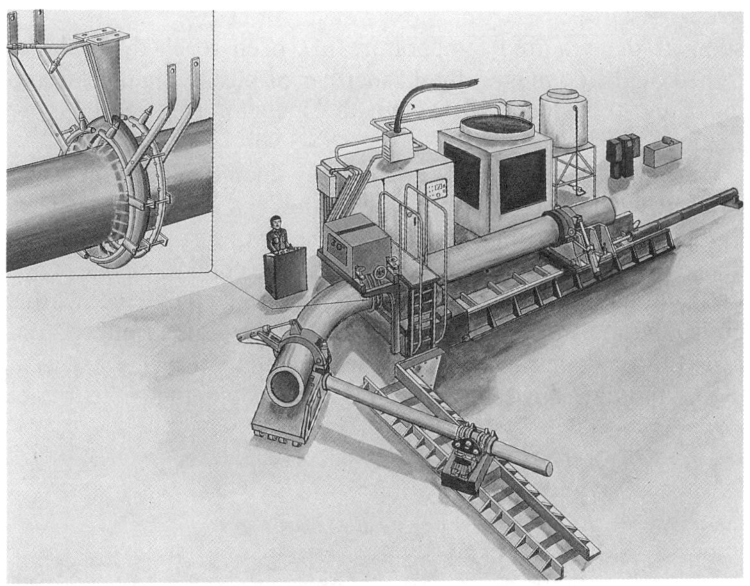

Figure 2-21. Gregson induction bender—basic design format. (Courtesy of Gregson Pipe Work Ltd.)

Figure 2-22. A closer view of the water spray. (Courtesy of Inkamaf Corp.)

soon as the required temperature has been reached, predetermined by the computer, the pusher bar or pusher clamp starts to move the pipe forward through the inductor ring. See Figure 2-23.

Because the metal is now sufficiently ductile at this point, and yet still nonductile on either side of the heat band, and because the front end is firmly clamped to the pivot arm, a bend is induced at the tangent point. The process is a continuing one, once initiated, until the full arc of the bend has been completed. At this point, the machine shuts off and the bent pipe is removed. Because in most instances it has been continuously cooled as it passes through the heated area, it is cool enough to handle imme-

Figure 2-23. Induction bend in progress. Note the narrow width of the heat band and the two pyrometers for monitoring the band temperature. (Courtesy of Inkamaf Corp.)

diately and only two men are needed to load, bend, and unload a piece of pipe of virtually any size.

The process is very much faster than the traditional hot slab method and far less labor intensive. Usually a pipe can be loaded, bent, and unloaded in less time than it takes to sand-fill the same size pipe preparatory to heating it in an oven, which may take several hours, before removing it and starting to make the bend.

In comparison to cold bending, however, it does take longer, especially for the smaller pipe diameters. Multiple bends in say 4″ Schedule 40 pipe can be made far more quickly by cold bending than by induction. However, when one takes into consideration certain other factors such as additional handling time for lubricant washout in cold bending, machine limitations because of wall thickness (a 4″ machine will not bend anything heavier than Schedule 80 carbon steel at 1½D, which would have to be done on a 6″ machine), tooling change downtime for radius changes in the same size material, and even wall thickness changes, then the time differences are not so significant. In fact, under certain conditions induction bending can actually be quicker.

There are limitations, though. Nonferrous material as yet cannot be bent by the induction method. This rules out aluminum, copper, and some of the copper nickel materials. It also rules out thin wall pipe in tight radius configuration, at least for the time being. Small induction benders are very new to the industry, and there is little doubt that both manufacturers and bending machine users will get into extensive R & D programs to tackle these particular aspects of bending by the induction method. See Figures 2-24 and 2-25.

One of the biggest advantages of some induction benders is the use of their computers. Cojafex, for example, provides a Hewlett Packard 85 desktop unit with pre-programmed tape cassette for its various benders.

An inquiry concerning a particular bend can be responded to immediately simply by keying in the pertinent information. The computer will very quickly assess this and reply with all needed details.

Figure 2-24. Offset induction bend. (Courtesy of Houston Pipe Benders.)

Figure 2-25. 1½ D small induction bender. The pivot point is adjustable to accommodate a range of pipe sizes from 2″–12″. (Courtesy of Inkamaf Corp. and International Piping Systems, Ltd.)

Let's see how this is done. The inquiry is as follows:

Material: 16″ standard wall, A-106, Grade B pipe
Bend: 90°-48″ CLR (center line radius)
Quantity: one only

This information is keyed into the computer and when completed, the screen shows a visual readout of all the bending information and data applicable. Additionally, with a machine operation cost factor already programmed into the system, the cost of making the bend is also shown.

In this particular case, we have used the cost factor of $250/hour and if the visual readout is satisfactory, the computer will print out a hard copy version.

For our particular bend, see the printout in Table 2-1A. Note that in addition to the pipe dimensions, i.e., OD, wall thickness, and bend data (Bend Factor R/D and Wall Factor OD/t), which for cold bending are normally determined from the tooling charts, all the critical information such as temperature, yield strength, pushing and bend roll forces, and induction power are also shown.

Two very important items—how much wall thinning and how much ovality—are also shown and this, of course, is one of the really big advantages induction bending has over other methods. The wall thinning and ovality are predictable. With an experienced operator at the controls these predicted figures may well be improved upon.

Finally, we have the bending real-time, the floor-to-floor real-time and the cost. One interesting item of information is the bending length. If one calculates the developed length of material for this particular radius and degree of bend, one will find that it is somewhat less than shown on the printout, i.e., calculated developed length is 75.3984″, whereas the printout shows 78.66″, which means that the amount of additional material *compressed* in order to make the bend with effective wall thinning is 3.2616″.

In this case, the wall thinning will not be more than 11.45% and the ovality no more than 3.91%. It will take just over half an

Table 2-1A
Cojafex PB 850 Pipe Bender Data

COJAFEX PB 850 PIPE BENDER		
Outside diam.	16.00	inch
Wall thickness	.38	inch
Bending radius	48.00	inch
Temperature	1742.00	deg. F
Bending angle	90.00	deg.
Material factor	1.30	
Number of bends	1.00	
Machine price	250.00	$/hr.
—		
R/D	3.00	
D/S	42.11	
Ovality	3.91	%
Yield strength	3629.61	PSI
Bending moment	.34	lbs.in.x10^6
Pushing force	.70	lbs. x10^4
Wall thinning	11.45	%
Bending length	78.66	inch
Min. pipe length	160.16	inch
Bend. roll force	1.34	lbs. x10^4
Bending speed	2.47	inch/min.
Induct. power	212.06	kW
Bending time	31.85	min.
Floor to floor	67.13	min.
Cost price	279.71	$

hour to bend. The whole operation—floor-to-floor time—will take an hour and 7 minutes, at a cost of $279.71.

Now let's say the client wants a tighter radius than 3D, but we don't wish to use the charts to determine this, since these are basically only guidelines. So we start at 1½D, and key that radius, i.e., 24″, into the computer, which very quickly says REJECTED. (Table 2-1B.) We can then start increasing the radius until the computer accepts a particular figure. Rather than using fractions of an inch, we use 2″ increments and get a response at 40″. Consequently, the material requirement is reduced as is the bending time and speed and inductive power, with increases, however, in ovality and wall thinning, because of a more severe bend radius. (Table 2-1C.)

Table 2-1B
Cojafex PB 850 Pipe Bender Data

COJAFEX PB 850 PIPE BENDER		
Outside diam.	16.00	inch
Wall thickness	.38	inch
Bending radius	24.00	inch
Temperature	1742.00	deg. F
Bending angle	90.00	deg.
Material factor	1.30	
Number of bends	1.00	
Machine price	250.00	$/hr.
—		
R/D	1.50	MINIMUM
D/S	42.11	REJECTED
Ovality	7.82	%
Yield strength	4210.91	PSI
Bending moment	.39	lbs.in.x10^6
Pushing force	1.63	lbs. x10^4
Wall thinning	20.99	%
Bending length	41.30	inch
Min. pipe length	122.80	inch
Bend. roll force	1.55	lbs. x10^4
Bending speed	1.13	inch/min.
Induct. power	155.65	kW
Bending time	36.40	min.
Floor to floor	71.68	min.
Cost price	298.65	$

Table 2-1C
Cojafex PB 850 Pipe Bender Data

COJAFEX PB 850 PIPE BENDER		
Outside diam.	16.00	inch
Wall thickness	.38	inch
Bending radius	40.00	inch
Temperature	1742.00	deg. F
Bending angle	90.00	deg.
Material factor	1.30	
Number of bends	1.00	
Machine price	250.00	$/hr.
—		
R/D	2.50	
D/S	42.11	
Ovality	4.69	%
Yield strength	3735.44	PSI
Bending moment	.35	lbs.in.x10^6
Pushing force	.87	lbs. x10^4
Wall thinning	13.43	%
Bending length	66.21	inch
Min. pipe length	147.71	inch
Bend. roll force	1.37	lbs. x10^4
Bending speed	2.20	inch/min.
Induct. power	200.78	kW
Bending time	30.06	min.
Floor to floor	65.34	min.
Cost price	272.24	$

Now, had a minimum radius been critical, the next alternative would have been to go to a heavier wall, say, Schedule 40 at .500″. Starting again at 1½D, we quickly arrive at a radius of 28″ or 1.75D as being acceptable for bending. (Table 2-1D).

However, while the time and cost have increased marginally, the wall thinning and ovality have increased significantly. If these are critical factors, further trial and error with the computer will produce an acceptable balance between minimum radius, wall thickness necessary, and achievable wall thinning and ovality. And, of course, we will always have the time and cost figures.

This computerized method of bending, along with the machine's unique capability to adjust the radius to suit critical factors, has a tremendous advantage over all other types of bending.

Table 2-1D
Cojafex PB 850 Pipe Bender Data

COJAFEX PB 850 PIPE BENDER		
Outside diam.	16.00	inch
Wall thickness	.50	inch
Bending radius	28.00	inch
Temperature	1742.00	deg. F
Bending angle	90.00	deg.
Material factor	1.30	
Number of bends	1.00	
Machine price	250.00	$/hr.
—		
R/D	1.75	
D/S	32.00	
Ovality	5.55	%
Yield strength	4614.70	PSI
Bending moment	.55	lbs.in.x10^6
Pushing force	1.98	lbs. x10^4
Wall thinning	18.35	%
Bending length	47.45	inch
Min. pipe length	128.95	inch
Bend. roll force	2.20	lbs. x10^4
Bending speed	1.68	inch/min.
Induct. power	178.61	kW
Bending time	28.28	min.
Floor to floor	64.24	min.
Cost price	267.66	$

Imagine what this would cost if applied to the smaller diameters of pipe at 12″ and under if one had to use cold bending methods. The tooling costs alone would be prohibitive.

So when we look at piping design from now on, it is easy to appreciate that designers and engineers will be far more flexible than they have been. Instead of being constrained to a basically inflexible 1½D configuration, they will be able to design to optimum radii and know exactly the wall thinning or ovality they can expect, which, at worst, (1½D) would be less than 12½% and proportionally better than that as the radii are increased. See Figures 2-26 and 2-27.

Very recently a Norwegian company has introduced two small induction benders—one for up to 4″ pipe, the other for up to 6″. Designed for the smaller shop or field application, the unique feature of these machines is that they bend vertically rather than

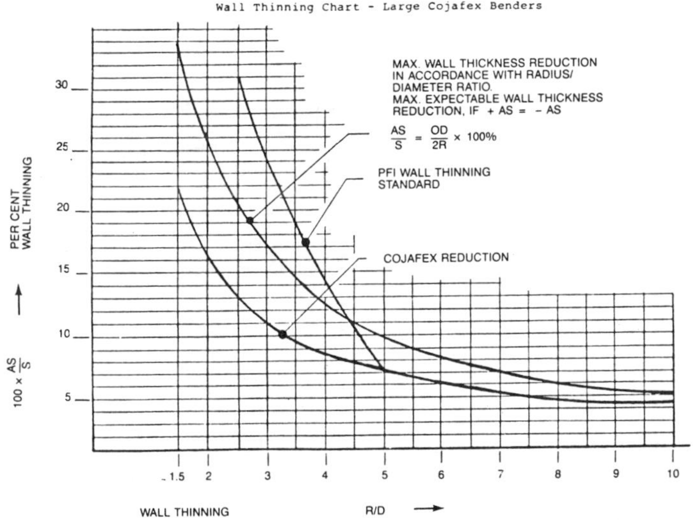

Figure 2-26. Wall thinning chart—Large Cojafex Benders.

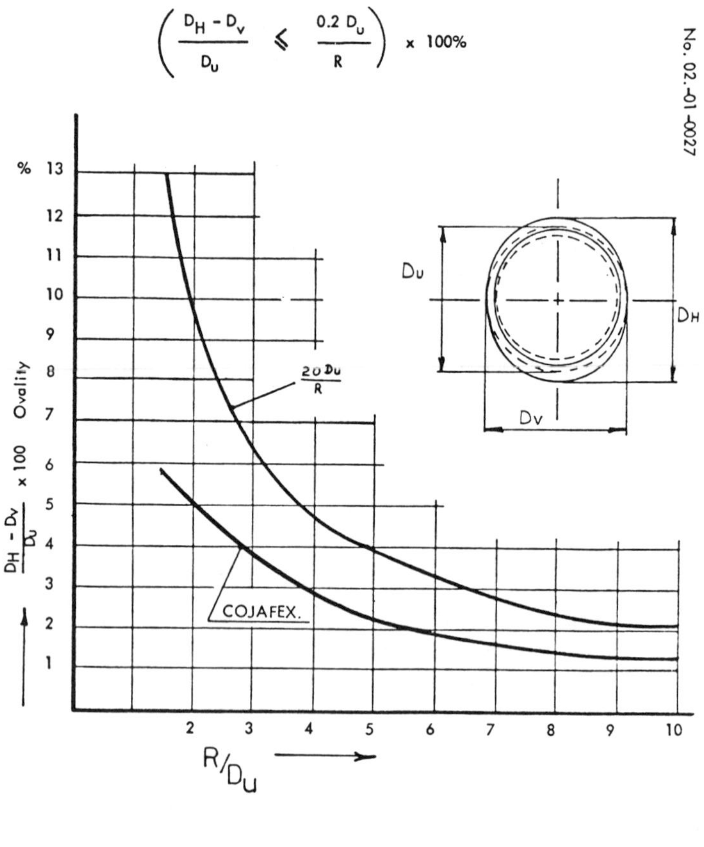

Figure 2-27. Allowed ovality according to "VGB-Dampftechnik GmbH, Essen."

on a horizontal plane. Floor space requirements, therefore, are reduced to a minimum (Figure 2-28).

Preliminary information (on the machines) tends to show that minimum radii are in the 3D area with capability oriented toward thinner wall pipe and tube rather than the heavier walls.

Significantly, this latest addition to the induction bending field indicates growing emphasis on the process particularly with respect to the smaller pipe and tube diameters, hitherto the domain of cold bending machinery.

Considerable research has been done on the effects induction bending has on the metallurgical structures of pipe. Table 2-2 il-

Figure 2-28. Ell-bend mobile pipe bending machine. (Courtesy of Goltens Engineering A/S.)

Table 2-2
Summary of Induction Bending of Various API-5LX Grade Pipe

Pipe grade Size-type PBHT	Yield Strength Before/After (psi)	Tensile Strength Before/After (psi)	Elongation Before/After (% in 2″)	Toughness Charpy "v" notch Before/After (ft-lbs)
Grade X42				
12¾″ OD × .500 wall (SMLS)	55,900/59,900	65,900/77,800	42.5/41.0	None required
Grade X52				
24″ OD × .438″ wall (Tempered-DSAW)	66,200/77,800	81,900/88,500	34.5/33.0	93/91 (@ −50°F.)
Grade X56				
10¾″ OD × .356 wall (SMLS)	75,500/88,100	85,500/109,400	31.0/29.5	None required
Grade X60				
24″ OD × .485 wall (Tempered-DSAW)	66,400/73,400	83,200/86,800	33.0/30.0	26/35 (@ +32°F.)
24″ OD × .688 wall (Tempered-DSAW)	61,000/73,700	80,500/86,400	39.0/40.5	35/52 (@ +32°F.)
26″ OD × .312 wall (DSAW)	67,100/68,300	86,100/97,700	29.0/25.5	18/22 (@ +32°F.)

30″ OD × :600 wall (DSAW)	63,500/70,200	81,200/98,100	40.0/33.0	42/48 (@ +32°F.)
30″ OD × .660 wall (DSAW)	61,200/66,100	81,900/99,500	41.0/36.0	41/41 (@ +32°F.)
30″ OD × .625 wall (DSAW)	64,600/72,300	84,100/99,300	36.5/30.5	15/34 (@ +32°F.)
Grade X65				
10¾″ OD × .550 wall (Tempered-SMLS)	77,300/77,200	91,500/93,000	34.0/31.0	171/88 (@ −50°F.)
16″ OD × .312 wall (Tempered-DSAW)	78,700/71,500	90,400/84,500	26.0/35.0	47/100 (@ −50°F.)
24″ OD × .344 wall (DSAW)	68,700/69,800	85,700/103,500	29.0/31.0	18/19 (@ +32°F.)
24″ OD × .438 wall (Tempered-DSAW)	74,600/75,500	90,700/90,000	29.0/30.0	89/113 (@ −50°F.)
30″ OD × .562 wall (Tempered-DSAW)	65,700/71,200	85,300/88,100	31.5/32.0	39/45 (@ +32°F.)
30″ OD × .750 wall (Tempered-DSAW)	65,000/71,000	90,900/89,400	37.0/26.0	71/147 (@ +32°F.)
Grade X70				
30″ OD × 1.313 wall (Tempered-DSAW)	78,800/73,600	88,600/85,200	27.5/29.0	150/284 (@ 0°F.)

Data supplied courtesy of Robert L. Hipley, Technical Services Manager, Associated Piping & Engineering Corp.

lustrates some of the findings, in respect to various grades of API-5LX pipe, arrived at by Associated Piping and Engineering. A similar study on X-65 and X-70 line pipe was conducted by Fabricom in Belgium and the findings published in the May 1983 issue of *Pipe Line Industry.* Another study by Dai-Ichi High Frequency Co. in Japan on the comparative strength of butt-weld elbows (fittings) and induction bends at 1½D began in 1975 with the findings presented at the Pipetech Conference in Houston in February 1984.

Undoubtedly, many more such studies will be conducted as this process becomes more widely used, particularly for the smaller diameters of material.

Hot Slab Bending

Hot slab bending is perhaps the oldest form of pipe bending in existence and is a method that is still widely used today, especially for large diameter and/or heavy wall pipe. The methodology is basically very simple. First the pipe is filled with completely dry silica sand. Some benders used to take regular wet beach sand, pack it in the pipe, close off the ends and then heat up the bend area with a torch. Result: the pipe exploded with often serious consequences to anyone near by. The moisture in the sand, when heated, turns to steam and has to expand. Since the area on either side is packed solid and the ends sealed, the pipe has to give in the heated area.

Hot slab bending shops fill their pipes in a vertical pit as much as 60 feet deep using mechanical rappers to assist in helping pack the sand down tight. The bending foreman's long experience will tell him when the sand is packed to the correct density. Often he can tell by slapping his hand on the surface of the material at the top.

When this stage has been completed the pipe is removed, and placed in an oven where the ends are bricked up to close around the pipe and contain the heat in the bend area. Then the oven is fired up—gas heating is the usual method—until the bend area reaches a temperature of somewhere between 2,000° and 2,100°F. As soon as the pipe is ready to be bent, it is removed from the oven and placed on a platform or bending table where one end is securely fastened in clamps and a cable is attached to the other. Shaped bending forms are then positioned at predetermined points while the cable end is gradually pulled in by a winch or other suitable method. See Figure 2-29.

It is a slow process requiring constant attention. Bending must be completed before the pipe cools to 1,600°F. Difficult and large-radius bends may have to be bent in two or three or more stages. It is not uncommon for a single bend of this nature to take a day or longer to achieve. Radii for hot slab bending are generally limited to a minimum of 3D to 5D, and are virtually impossible below this without buckling or wrinkling the pipe.

Even now, with induction machines capable of bending 70″ or larger diameter pipe with wall thicknesses of up to 4″ or more, anything that cannot be handled by induction must be done by the hot slab method.

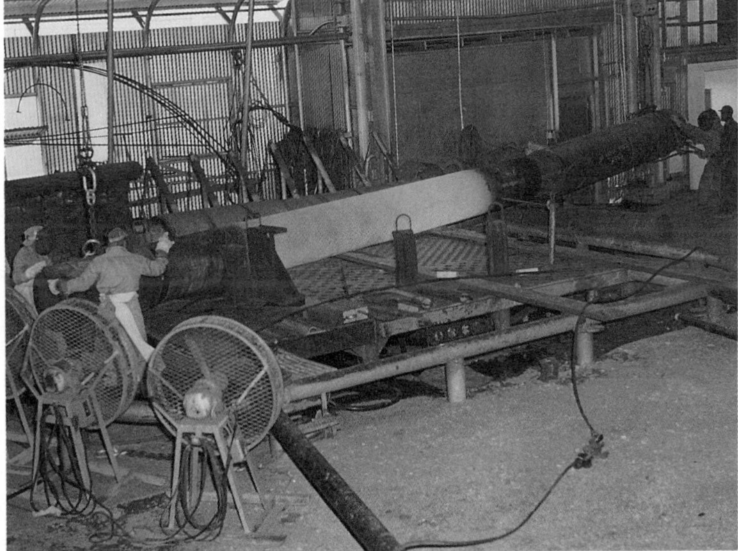

Figure 2-29. Pipe being positioned on the bending table. The end at the left will be clamped while the other end will be pulled by wire cable against the forming blocks. (Courtesy of Texas Pipebenders.)

However, because of the higher temperatures needed for the hot slab process, some materials cannot be bent if they cannot withstand temperatures beyond 2,000°. In the same way, some materials cannot be bent by cold process if they are too large or have too heavy a wall. In a number of cases, however, induction bending has been able to resolve this problem, providing that the wall thickness is within the bend parameters, since the bending temperatures are not only lower than for hot slab bending, but are applied only to a narrow band of the pipe for a relatively short period of time. See Tables 2-3 and 2-4.

Table 2-3
Manufacturer A—Typical Parameters for Induction Bends Showing Minimum Achievable Radii in Respect to Pipe OD and Wall Thickness

WALLTHICKNESS / OUTSIDE DIAMETER	3"	4"	6"	8"	10"	12"	14"	16"	18"	20"	22"	24"	26"	28"	30"
	88,9	114,2	168,3	219,4	273	323	355,8	406,4	457,2	509	558,8	609,6	660,4	711,2	762
3	265	440	950	1590	2000	2500	2800	3400	4000	4700	5500	6700	8800	13400	20000
4	200	330	715	1195	1530	1895	2195	2660	3195	3805	4500	5285	6000	9000	15290
5	180	265	570	950	1220	1505	1755	2125	2555	3040	3600	4220	5260	8600	11610
6	180	230	480	795	1020	1260	1465	1770	2130	2535	3000	3520	4120	5470	8960
7	180	200	410	680	875	1080	1255	1520	1820	2170	2530	3000	3530	4900	7000
8	180	180	290	500	750	940	1100	1325	1560	1900	2250	2540	3090	4210	5500
9	180	180	280	500	680	840	915	1180	1420	1690	2000	2345	2750	3640	4660
10	180	180	280	500	600	740	850	1050	1270	1500	1790	2110	2470	3250	4100
11	180	180	260	500	540	680	790	950	1155	1375	1630	1920	2250	3000	3770
12	180	180	260	500	500	630	730	885	1065	1270	1500	1760	2060	2600	3280
13	180	180	260	500	500	545	660	815	980	1170	1385	1625	1900	2270	2850
14	180	180	260	500	500	510	610	750	900	1070	1285	1490	1745	2140	2610
15	180	180	260	500	500	500	540	685	815	970	1180	1350	1580	1945	2320
16	180	180	260	500	500	500	540	620	730	875	1085	1220	1425	1755	2090
17		180	260	500	500	500	540	620	710	780	985	1085	1270	1470	1890
18		180	260	400	420	490	540	620	690	780	880	1010	1180	1380	1725
19		180	260	350	420	490	540	620	690	780	870	975	1140	1340	1580
20		180	260	340	420	490	540	620	690	780	860	950	1100	1300	1450
22			260	330	420	490	540	620	690	770	850	950	1010	1150	1320
24			260	330	420	490	540	620	690	770	850	950	1010	1100	1210
26			260	330	420	490	540	620	690	770	850	950	1010	1100	1160
28			260	330	420	490	540	620	690	770	850	950	1010	1100	1150
30			260	330	420	490	540	620	690	770	850	950	1010	1100	1150
35				330	420	490	540	620	690	770	850	950	1010	1100	1220
40				330	420	490	540	620	690	770	850	950	1010	1100	1260
45				330	420	490	540	620	690	770	850	950	1010	1100	1295
50				330	420	490	540	620	690	770	850	950	1010	1100	1315
55					420	490	540	620	690	770	850	950	1010	1100	1335
60					420	490	540	620	690	770	850	950	1180	1240	1370
65					420	490	540	620	690	770	850	950	1240	1260	1390
70					420	490	540	620	690	770	850	950	1255	1310	1410
75					420	490	540	620	690	770	850	950	1290	1350	1450
80					420	490	540	620	690	770	850	950	1300	1355	1455
85						490	540	620	690	770	850	950	1320	1385	1460
90						490	540	620	690	770	850	950	1330	1395	1485
95						490	540	620	690	770	850	950	1335	1405	1500
100						490	540	620	690	770	850	950	1380	1420	

Table 2-4
Manufacturer B—Typical Parameters for Induction Bends Showing Minimum Achievable Radii in Respect to Pipe OD and Wall Thickness

OUTSIDE DIAMETER (INCHES/MILLIMETRES)

WALL THICKNESS (INCHES/MILLIMETRES)	3½" / 88.9	4½" / 114.3	5½" / 139.7	6⅝" / 168.2	7⅝" / 193.6	8⅝" / 219	9⅝" / 244.4	10¾" / 273	12¾" / 323.8	14" / 355.6	16" / 406.4	18" / 457.2	20" / 508	22" / 558.8	24" / 609.6	26" / 660.4	28" / 711.2	30" / 762
.192" / 4.87	114	152	191	259	353	461	585	721	1039	1415	1849	2340	2890	3498	4163	4880	5661	6499
.212" / 5.38	114	152	191	234	320	418	528	652	941	1280	1674	2116	2616	3168	3767	4424	5127	5890
.25" / 6.35	114	152	191	228	272	353	448	553	798	1088	1418	1796	2219	2682	3194	3751	4352	4991
.281" / 7.13	114	152	191	228	266	314	400	492	710	967	1263	1600	1971	2386	2840	3335	3868	4442
.312" / 7.92	114	152	191	228	266	305	358	444	640	871	1137	1440	1778	2151	2560	3004	3484	4000
.375" / 9.52	114	152	191	228	266	305	342	381	530	725	946	1179	1478	1788	2127	2496	2901	3329
.438" / 11.12	114	152	191	228	266	305	342	381	457	618	808	1024	1264	1531	1822	2139	2482	2849
.5" / 12.70	114	152	191	228	266	305	342	381	457	544	707	896	1107	1341	1597	1875	2176	2491
.563" / 14.30	114	152	191	228	266	305	342	381	457	533	629	795	985	1190	1414	1664	1927	2217
.625" / 15.87	114	152	191	228	266	305	342	381	457	533	609	717	883	1072	1274	1499	1735	2110
.6875" / 17.46		152	191	228	266	305	342	381	457	533	609	685	802	972	1158	1360	1742	2308
.75" / 19.05		152	191	228	266	306	342	381	457	533	609	685	762	894	1060	1393	1884	2499
.8125" / 20.63		152	191	228	266	305	342	381	457	533	609	685	762	838	1078	1499	2026	2712
.875" / 22.22				228	266	305	342	381	457	533	609	685	762	838	1152	1598	2169	2880
.9375" / 23.81				228	266	305	342	381	457	533	609	685	762	854	1225	1703	2311	3070
1.0" / 25.40				228	266	305	342	381	457	533	609	685	762	905	1298	1802	2446	3246
1.0625" / 26.98				228	266	305	342	381	457	533	609	685	762	949	1365	1901	2581	3429
1.125" / 28.57				228	266	305	342	381	457	533	609	685	762	1000	1438	2001	2716	3611
1.1875" / 30.16				228	266	305	342	381	457	533	609	685	762	1044	1505	2093	2844	3787
1.25" / 31.75						305	342	381	457	533	609	685	762	1089	1572	2192	2979	3962
1.312" / 33.29						305	342	381	457	533	609	685	762	1134	1633	2284	3107	4130
1.37" / 34.79						305	342	381	457	533	609	685	787	1179	1694	2364	3221	4290
1.43" / 36.32						305	342	381	457	533	609	685	817	1218	1755	2456	3342	4450
1.50" / 38.10						305	342	381	457	533	609	685	848	1268	1828	2555	3477	4632
1.75" / 44.45						305	342	381	457	533	609	685	949	1424	2066	2892	3947	5273
2.0" / 50.80						305	342	381	457	533	609	685	1046	1575	2286	3209	4395	5875
2.25" / 57.15								381	457	533	609	713	1132	1709	2493	3506	4807	6446
2.5" / 63.50								381	457	533	609	763	1214	1838	2682	3784	5198	6979

MINIMUM RADIUS IN MILLIMETRES.

The hot slab process is slow and highly labor intensive in comparison to cold and induction bending, but it is still the only way to bend certain categories of pipe. Although technology is constantly improving in all areas of mechanical bending, the hot slab method will be around for quite some time to come. Unfortunately, particularly for the larger pipe diameters, certain bending standards and tolerances are still based on the hot slab process, despite the fact that these have been outdated and substantially improved upon by both the induction and cold processes, methods which have been in use for many years. Hopefully, this situation will change soon as the matter is now receiving growing attention from within the industry itself.

Perhaps we can look forward to seeing some national standards developed and established for *all* technologies and methods, not just the oldest one, within the next year or two.

3

Specialized Types of Bending

Aluminum Bending

Although aluminum pipe is virtually identical to carbon steel pipe in terms of nominal size and wall thicknesses, the similarity stops there insofar as bending is concerned, because of the characteristics of the material itself.

The first marked difference is of course in weight, aluminum being very much lighter and consequently easier to handle. 4″ Schedule 40 aluminum pipe, for example, weighs approximately 3.733 lbs/ft as opposed to 4″ Schedule 40 carbon steel or stainless steel, which weighs 10.79 lbs/ft—almost three times as much. A 20′ length of 4″ aluminum pipe weighs about 75 lbs, but the same length and size in carbon steel would weigh over 200 lbs.

That's only a minor detail, however, The real problem in bending is due to the various tempers of the aluminum alloys. Most aluminum pipe used, especially for air conveying purposes, is in 6061 or 6063 alloy and T6 temper, which is quite hard. Also, the material continues to age-harden after production and so while it is possible to bend T6 material, the older it is the tougher it is to bend. In fact, quite often it will break rather than bend, particularly if it has been in storage for any length of time. It also has a horrendous amount of springback.

New 4″ Schedule 40 T6 pipe, for example, for a 42″ radius would probably need a 36″ centerline radius die—in other words, it has something like 6″ of springback on that radius. If the material has been in storage for some time, the chances are it won't even bend. And one cannot use the same radius die for the required radius bend—it must have reduction in the actual radius of the die itself for radial growth allowance.

Generally, there are two approaches to this problem. One is to have the pipe annealed down to T0 and then bend it, without need for springback allowance. The other is to purchase the pipe in quantity in T4 condition. This latter temper is very bendable and the straight sections at each end are still machineable, i.e., can accept a Victaulic groove, whereas T0 material is too soft to be grooved. T0 material is also highly susceptible to damage in transit or through careless handling, particularly the thin-wall pipe. Nothing is more aggravating to a bender than to draw a length for bending only to find it is dented somewhere, and will not accept the mandrel. Sometimes the dent can be pushed out by the mandrel if it is in T0 condition, but more often than not the piece must be rejected.

In addition to these problems, even though the cold bending process work hardens the pipe, it still does not bring it back to full T6 condition, particularly if it started out in T0. T4 on the other hand, work-hardened, comes closer to T6, but is still not quite T6 in the bend area. T0 is very definitely somewhat harder in the bend area and very much still T0 in the tangents.

Reynolds Aluminum has charted minimum radii for bending T6 aluminum pipe, but this is conditional upon using fully tooled radial draw equipment, and (presumably) brand new, straight-from-the-mill material and bending virtually the moment it arrives—ideally the same day it was produced. See Figure 3-1.

The aluminum alloys, both non-heat treatable and heat treatable, for cold working purposes, are divided into four classes: A, B, C, and D. And where maximum formability is required,

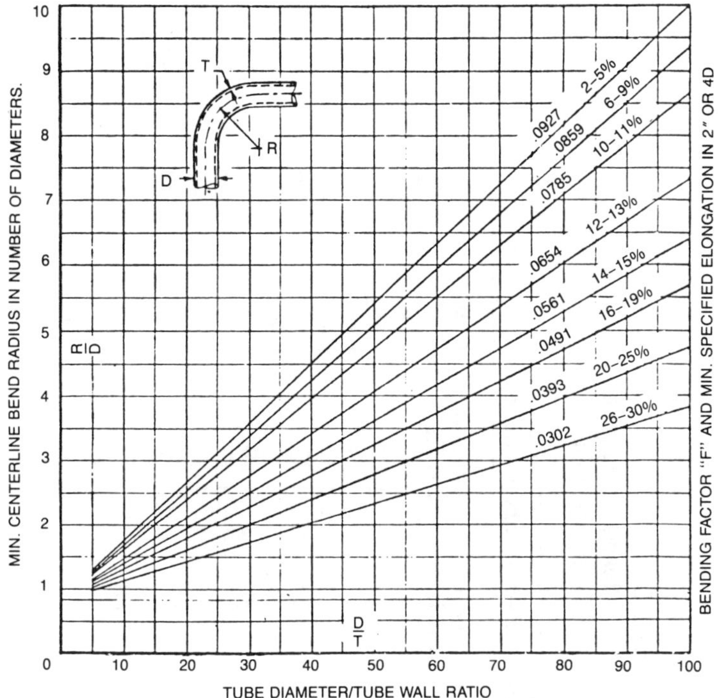

Figure 3-1. 1) Elongation figures are minimum specified as listed in the Aluminum Association's Standards and Data handbook. 2) Limitations are based on tooled and boosted radial draw bending machinery.

1060, 1100, and 3003 of the non-heat treatable and 6061 and 6063 of the heat treatable alloys give the best results.

Table 3-1 of eight formability groups based on elongation properties will provide a quick reference for determining minimum radii achievable. Caution is advised, however, in using this table, as one can *never* be 100% certain of the material's age *or* its bendability until one actually starts bending. Even with the best possible equipment and under ideal circumstances and conditions, one can be surprised, from time to time.

Table 3-1
Radial Draw Bending—Bending Limitations
Approximate Minimum Centerline Bend Radii

Diameter Wall Ratio $\frac{D}{T}$	Minimum Centerline Bend Radii (in number of tube diameters)							
	Minimum Specified Elongation in 2″ or 4D.							
	26–30%	20–25%	16–19%	14–15%	12–13%	10–11%	6–9%	2–5%
5	.94	.98	1.03	1.07	1.11	1.18	1.21	1.25
6	.97	1.02	1.08	1.12	1.18	1.26	1.30	1.34
7	.10	1.06	1.13	1.18	1.24	1.33	1.39	1.43
8	1.03	1.10	1.18	1.23	1.31	1.41	1.47	1.53
9	1.06	1.14	1.23	1.29	1.37	1.49	1.56	1.62
10	1.09	1.18	1.28	1.35	1.44	1.57	1.64	1.71
11	1.12	1.22	1.33	1.40	1.50	1.65	1.73	1.80
12	1.15	1.26	1.37	1.46	1.57	1.73	1.82	1.90
13	1.18	1.30	1.42	1.51	1.63	1.81	1.90	1.99
14	1.21	1.34	1.47	1.57	1.70	1.89	1.99	2.08
15	1.24	1.37	1.52	1.63	1.77	1.96	2.07	2.18
16	1.27	1.41	1.57	1.68	1.83	2.04	2.16	2.27
17	1.30	1.45	1.62	1.74	1.90	2.12	2.25	2.36
18	1.33	1.49	1.67	1.79	1.96	2.20	2.33	2.45
19	1.36	1.53	1.72	1.85	2.03	2.28	2.42	2.55
20	1.39	1.57	1.77	1.91	2.09	2.36	2.50	2.64
21	1.42	1.61	1.82	1.96	2.16	2.43	2.59	2.73

22	1.45	1.65	1.87	2.02	2.22	2.51	2.67	2.82
23	1.48	1.69	1.91	2.08	2.29	2.59	2.76	2.92
24	1.51	1.73	1.96	2.13	2.35	2.67	2.85	3.01
25	1.54	1.77	2.01	2.19	2.42	2.75	2.93	3.10
30	1.69	1.96	2.26	2.47	2.75	3.14	3.36	3.57
35	1.84	2.16	2.50	2.75	3.07	3.53	3.79	4.03
40	1.99	2.36	2.75	3.03	3.40	3.93	4.22	4.49
45	2.14	2.55	2.99	3.31	3.73	4.32	4.65	4.96
50	2.30	2.75	3.24	3.59	4.06	4.71	5.08	5.42
55	2.45	2.95	3.49	3.87	4.38	5.10	5.51	5.88
60	2.60	3.14	3.73	4.15	4.71	5.50	5.94	6.35
65	2.75	3.34	3.98	4.43	5.04	5.89	6.37	6.81
70	2.90	3.54	4.22	4.71	5.36	6.28	6.80	7.27
75	3.05	3.73	4.47	4.99	5.69	6.67	7.23	7.74
80	3.20	3.93	4.71	5.27	6.02	7.07	7.66	8.20
85	3.35	4.13	4.96	5.55	6.34	7.46	8.09	8.66
90	3.50	4.32	5.20	5.83	6.67	7.85	8.52	9.13
95	3.65	4.52	5.45	6.11	7.00	8.24	8.95	9.59
100	3.81	4.72	5.70	6.39	7.32	8.64	9.38	10.05

1. Limitations are based on fully tooled and boosted radial draw bending machinery.
2. Example: Find minimum centerline bending radius for 2″ diameter by .083″ wall, 6061-T6 tube. Divide diameter by wall thickness (2 ÷ .083 = 24). In line with diameter/wall ratio column for ratio 24 under elongation subgroup 10–11%, read 2.67. Multiply figure with diameter or 2 × 2.67 = 5.34″. In fractions, round to 5 3/8″.
3. If slight deformation can be tolerated in the bending zone, minimum centerline radius can be reduced by 1/4 tube diameter when inside forming mandrel is omitted. When press bending with fully supporting wing type tooling and/or compression bending using the follow block method, multiply values by 2. However, minimum centerline radius to be not less than 3D. All other bending methods, multiply values by 2.5 or no less than 4D.

Bending Aluminum Tubes and Shapes

General Bending Limitations

The limitations presented in Tables 3-2, 3-3, 3-4, and 3-5 are based on fully tooled and boosted cold draw bending at normal room temperature. Since the ductility of aluminum increases at elevated temperatures, hot bending may be employed in some cases. However, since mechanical properties are lowered, hot bending does not eliminate such common defects as buckling or collapsing. Temperatures and reheating times as indicated in the chart should not be exceeded.

When designing components of aluminum tubes and shapes, the largest practical bending radii should be selected regardless of indicated minimum limits. Borderline situations usually result in higher equipment, tooling, and maintenance costs as well as increased scrap.

When bending tubes and shapes of low-strength alloys to small centerline radii, springback is often negligible. Bending of larger radii and/or stronger alloys generally results in varying degrees of springback which must be compensated for by over-bending and/or cut-back design of tooling.

Aluminum Bending Formulas

We know that when a pipe or tube is bent the extrados thins and the intrados thickens, displacing the neutral axis inwardly to a new center of gravity.

We also know that maximum stretch and compression in the bend area have been reached at about 20°–30° into the bend and that maximum stresses have been reached at about the 90° point.

(text continued on page 74)

Table 3-2
Reynolds Standard Welded Tube
Bending Limitations
Approximate Minimum Centerline Bend Radii
Alloys 3005-H28, H294, & 3005-H29, H295

Outside Dimension (in.)	Wall Thickness (in.)												
	.025	.027	.029	.031	.033	.035	.037	.039	.041	.047	.057	.063	.071
1/2 OD	1.25	1.16	1.11	1.07	1.03	1.00	.95	.93	.90	.84	.76	.72	.69
5/8 OD	1.85	1.75	1.66	1.59	1.52	1.46	1.41	1.36	1.32	1.21	1.00	1.03	.97
3/4 OD	2.51	2.37	2.25	2.14	2.04	1.96	1.89	1.82	1.76	1.61	1.43	1.35	1.27
7/8 OD	3.33	3.13	2.97	2.81	2.69	2.58	2.47	2.38	2.30	2.09	1.85	1.74	1.62
1 OD	4.23	3.98	3.75	3.56	3.40	3.24	3.11	3.00	2.89	2.62	2.30	2.16	2.00
1 1/8 OD	5.25	4.93	4.64	4.41	4.19	4.00	3.84	3.68	3.55	3.21	2.80	2.62	2.43
1 1/4 OD	6.34	5.94	5.60	5.30	5.04	4.81	4.60	4.42	4.25	3.83	3.33	3.11	2.87
1 3/8 OD	7.56	7.08	6.67	6.31	5.99	5.71	5.46	5.23	5.03	4.53	3.92	3.65	3.36
1 1/2 OD	8.9	8.33	7.84	7.41	7.03	6.69	6.40	6.13	5.89	5.29	4.57	4.24	3.90
2 OD	15.3	14.3	13.4	12.6	12.0	11.4	10.8	10.4	9.94	8.87	7.59	7.01	6.40
13/16 Sq	4.45	4.18	3.95	3.75	3.57	3.41	3.27	3.15	3.03	2.75	2.41	2.25	2.09
7/8 Sq	5.12	4.80	4.53	4.29	4.09	3.90	3.74	3.59	3.46	3.13	2.73	2.56	2.37
1 Sq	6.56	6.15	5.79	5.48	5.21	4.97	4.76	4.56	4.39	3.96	3.44	3.21	2.96
1 1/4 Sq	9.93	9.28	8.73	8.25	7.82	7.45	7.11	6.81	6.54	5.87	5.06	4.69	4.31
3/4 × 1 1/4	3.87	3.64	3.44	3.27	3.11	2.98	2.86	2.75	2.65	2.41	2.12	1.99	1.85
	9.93	9.28	8.73	8.25	7.82	7.45	7.11	6.81	6.54	5.87	5.06	4.69	4.31

Note: When calculating minimum centerline bend radii for tube sizes not listed, use bending Factor "F" = .0859. (For Alloy 3005-H25, use "F" = .0785.)

Table 3-3
Approximate Minimum Centerline Bend Radii
Radial Draw Bending of Round Tubing

Tube Diameter	Wall	Minimum Specified Elongation in 2″ or 4D 26–30%	20–25%	16–19%	14–15%	12–13%	10–11%	6–9%	2–5%
	.022	.28	.31	.33	.35	.38	.41	.43	.45
1/4	.028	.26	.28	.31	.32	.34	.37	.39	.41
	.035	.25	.27	.28	.30	.31	.33	.35	.36
	.028	.44	.49	.53	.56	.61	.68	.71	.75
3/8	.035	.42	.46	.50	.53	.56	.62	.65	.68
	.049	.38	.41	.44	.46	.49	.53	.55	.57
	.035	.61	.67	.74	.79	.85	.94	1.00	1.04
1/2	.049	.55	.59	.64	.68	.72	.79	.82	.86
	.065	.52	.55	.59	.62	.66	.71	.74	.77
	.049	.93	1.03	1.14	1.22	1.33	1.47	1.55	1.64
3/4	.065	.86	.95	1.03	1.10	1.18	1.30	1.37	1.43
	.083	.80	.86	.92	.97	1.03	1.12	1.17	1.22
	.049	1.16	1.30	1.46	1.57	1.72	1.93	2.04	2.14
7/8	.065	1.03	1.14	1.24	1.32	1.43	1.58	1.66	1.74
	.083	.98	1.07	1.16	1.23	1.31	1.44	1.51	1.58
	.049	1.39	1.57	1.77	1.91	2.09	2.36	2.50	2.64
	.065	1.24	1.37	1.52	1.63	1.77	1.96	2.07	2.18
1	.083	1.15	1.26	1.37	1.46	1.57	1.73	1.82	1.90
	.109	1.09	1.18	1.28	1.35	1.44	1.57	1.64	1.71
	.049	1.93	2.21	2.51	2.74	3.03	3.44	3.66	3.88
	.065	1.70	1.91	2.15	2.31	2.54	2.85	3.03	3.19

1 1/4	.083	1.55	1.71	1.90	2.04	2.21	2.45	2.59	2.73
	.109	1.40	1.53	1.66	1.75	1.88	2.08	2.16	2.25
	.065	2.22	2.54	2.87	3.12	3.44	3.89	4.14	4.38
	.083	2.00	2.24	2.50	2.69	2.94	3.30	3.50	3.68
1 1/2	.109	1.82	2.01	2.21	2.36	2.55	2.82	2.99	3.12
	.120	1.77	1.95	2.13	2.27	2.45	2.72	2.85	2.99
	.065	2.96	3.43	3.96	4.32	4.81	5.50	5.88	6.25
	.083	2.48	2.82	3.19	3.43	3.78	4.25	4.53	4.78
1 3/4	.109	2.22	2.47	2.75	2.94	3.20	3.57	3.78	3.97
	.120	2.17	2.40	2.66	2.85	3.10	3.43	3.62	3.81
	.065	3.38	3.92	4.52	4.94	5.50	6.28	6.72	7.14
	.083	3.02	3.46	3.92	4.26	4.70	5.34	5.70	6.02
2	.109	2.72	3.06	3.44	3.70	4.06	4.56	4.84	5.10
	.120	2.60	2.90	3.24	3.48	3.80	4.24	4.50	4.72
	.065	4.98	5.90	6.88	7.58	8.50	9.83	10.55	11.23
	.083	4.23	4.90	5.65	6.18	6.88	7.85	8.40	8.92
2 1/2	.109	3.70	4.23	4.78	5.20	5.73	6.48	6.90	7.30
	.120	3.55	4.03	4.55	4.90	5.40	6.08	6.48	6.83
	.065	6.42	7.65	8.97	9.93	11.19	12.96	13.95	14.88
	.083	5.52	6.48	7.50	8.25	9.21	10.59	11.37	12.09
3	.109	5.07	5.88	6.78	7.41	8.25	9.42	10.08	10.71
	.120	4.62	5.31	6.03	6.57	7.26	8.25	8.79	9.20

1. Above limitations based on fully tooled and boosted radial draw bending machinery.
2. When converting to fractional dimensions, reduce or increase to nearest 1/8″.
3. If slight deformation can be tolerated in the bending zone, minimum centerline radius can be reduced by 1/4 tube diameter when inside forming mandrel is omitted.
4. When press bending with fully supporting wing type tooling and/or compression bending using the follow block method, multiply values by 2. However, minimum centerline radius to be not less than 3D. All other bending methods, multiply values by 2.5 or no less than 4D.

Table 3-4
Welded Tube Bending Limitations
Approximate Minimum Centerline Bend Radii
Alloys 3005-H28, H294, & 3005-H29, H295

Outside Dimension (in.)	Wall Thickness (in.)												
	.025	.027	.029	.031	.033	.035	.037	.039	.041	.047	.057	.063	.071
1/2 OD	1.25	1.16	1.11	1.07	1.03	1.00	.95	.93	.90	.84	.76	.72	.69
5/8 OD	1.85	1.75	1.66	1.59	1.52	1.46	1.41	1.36	1.32	1.21	1.00	1.03	.97
3/4 OD	2.51	2.37	2.25	2.14	2.04	1.96	1.89	1.82	1.76	1.61	1.43	1.35	1.27
7/8 OD	3.33	3.13	2.97	2.81	2.69	2.58	2.47	2.38	2.30	2.09	1.85	1.74	1.62
1 OD	4.23	3.98	3.75	3.56	3.40	3.24	3.11	3.00	2.89	2.62	2.30	2.16	2.00
1 1/8 OD	5.25	4.93	4.64	4.41	4.19	4.00	3.84	3.68	3.55	3.21	2.80	2.62	2.43
1 1/4 OD	6.34	5.94	5.60	5.30	5.04	4.81	4.60	4.42	4.25	3.83	3.33	3.11	2.87
1 3/8 OD	7.56	7.08	6.67	6.31	5.99	5.71	5.46	5.23	5.03	4.53	3.92	3.65	3.36
1 1/2 OD	8.9	8.33	7.84	7.41	7.03	6.69	6.40	6.13	5.89	5.29	4.57	4.24	3.90
2 OD	15.3	14.3	13.4	12.6	12.0	11.4	10.8	10.4	9.94	8.87	7.59	7.01	6.40
13/16 Sq	4.45	4.18	3.95	3.75	3.57	3.41	3.27	3.15	3.03	2.75	2.41	2.25	2.09
7/8 Sq	5.12	4.80	4.53	4.29	4.09	3.90	3.74	3.59	3.46	3.13	2.73	2.56	2.37
1 Sq	6.56	6.15	5.79	5.48	5.21	4.97	4.76	4.56	4.39	3.96	3.44	3.21	2.96
1 1/4 Sq	9.93	9.28	8.73	8.25	7.82	7.45	7.11	6.81	6.54	5.87	5.06	4.69	4.31
3/4 × 1 1/4	3.87	3.64	3.44	3.27	3.11	2.98	2.86	2.75	2.65	2.41	2.12	1.99	1.85
	9.93	9.28	8.73	8.25	7.82	7.45	7.11	6.81	6.54	5.87	5.06	4.69	4.31

Note: When calculating minimum centerline bend radii for tube sizes not listed, use bending Factor "F" = .0859. (For Alloy 3005-H25, use "F" = .0785.)

Table 3-5
Aluminum Pipe, Alloy 6061-T6 and 6063-T6
Approximate Minimum Centerline Bend Radii

Nom. Pipe Size	Outside Diam. (in.)	Pipe Schedule Number 5	10	20	30	40	60	80	100	120	140	160	Mod. Wall
1/8	.405					.5		.44					
1/4	.540					.68		.62					
3/8	.675					.93		.81					
1/2	.840	1.5	1.31			1.12		1				.93	
3/4	1.050	2.12	1.81			1.56		1.31				1.2	
1	1.315	3.2	2.2			2		1.75				1.56	2.75
1 1/4	1.660	4.5	3.2			2.75		2.37				2.12	
1 1/2	1.900	5.68	3.93			3.31		2.81				2.43	
2	2.375	8.37	5.75			4.62		3.75				3	6.56
2 1/2	2.875	9.75	7.56			5.25		4.43				3.87	
3	3.500	13.87	10.37			6.93		5.75				4.75	10.06
3 1/2	4.000	17.62	13.12			8.37		6.81					
4	4.500	21.93	16.2			9.87		7.93		6.93		6.31	15.68
5	5.563	25.75	21.75			13.31		10.43		8.87		8	23
6	6.625	35.56	29.87			16.87		12.68		10.93		9.62	27.2
8	8.625	58.31	44.62	29.12	26.87	24	20.37	17.81	16	14.37	13.43	12.75	38.75
10	10.750	73.56	61.25	43.2	36.68	32.12	25.62	22.87	20.31				38.93
12	12.750	84.37	78.12	58.93	46.93	40	34.25	27.56					

1. Above limitations based on fully tooled and boosted radial draw bending machinery. When converting to fractional dimensions, reduce or increase to nearest 1/8″.
2. If slight deformation can be tolerated in the bending zone, minimum centerline radius can be reduced by 1/4 tube diameter when inside forming mandrel is omitted. When press bending with fully supporting wing type tooling and/or compression bending using the follow block method, multiply values by 2. However, minimum centerline radius to be not less than 3D. All other bending methods, multiply values by 2.5 or no less than 4D.

Also, that bending beyond 90° does not create any greater stresses.

To determine the minimum achievable centerline radius of bend for aluminum pipe or tube, one can use the following formula in application to draw bending in a fully-tooled condition:

$$R = \frac{D^2\pi}{4tS} + \frac{D\pi}{4}$$

or

$$R = .785D\left(\frac{D}{ts} + 1\right)$$

where R = centerline radius
D = outside diameter of the pipe or tube (*not* nominal diameter)
S = stretch (minimum specified elongation as listed in the Aluminum Association Standards Data handbook)
t = wall thickness

Tables 3-2 to 3-5 may be helpful in determining suitable radii and wall thicknesses for aluminum pipe and tube bending.

The various alloys and tempers are divided into eight elongation subgroups as shown in Table 3-6. The bending factors F for six of these subgroups are taken directly from the basic formula but modified slightly for the two subgroups covering the hardest tempers and elongation values of less than 9%. Since bending factor F is based on the lowest value of elongation range for each subgroup, a more accurate calculation may be made by using the basic formula directly on tube material with elongation values above 9%. Using the basic formula shown above where S = minimum specified elongation,

$$F = \frac{\pi}{4S}$$

Table 3-6
Application of Bending Formula
Bending Factor "F" and Elongation Subgroups

Tube Alloy & Temper	Bending Factor "F"	Elongation Subgroup
1060-0, 1060-H112, 1100-0, 3003-0	.0302	26–30%
1100-0, 3003-0, 5052-0	.0393	20–25%
5052-0, 6061-0, 6061-T4, 6063-0, 6063-T4	.0491	16–15%
2219-T31, T35 & T3511; 5083-0; 5086-0; 5454-0; 5456-0; 6066-T4; 2014-T4; 2024-T3	.0561	14–15%
7075-0, 5083-H111, 5086-H111, 5454-H111, 5456-H111, 6063-T1, 2024-0, 2014-0, 2014-T4	.0654	12–13%
6061-T6, 6063-T6, 1100-H12, 3003-H12, 5052-H32, 3005-H25	.0785	10–11%
1100-H14, H16; 3003-H14, H16; 6063-T832; 5052-H34, H36; 3005-H28, H29, H295; 3003-H28, H294	.0859	6–9%
1100-H18, 3003-H18, 5052-H38	.0927	2–5%

Formulas for Minimum Centerline Bend Radii (based on Fully Tooled Radial Draw Bender):

Round Tube

$$R = \frac{D^2F}{t} + .785\ D$$

Square Tube

$$R = \frac{E^2F}{t} + S$$

R = Centerline radius
D = Tube outside diameter
t = Wall thickness
F = Bending factor as listed above
S = Specified side dimension of square tube

E = Equivalent diameter $\left(\frac{4S}{3.14}\right)$

Aluminum Bending at Temperatures Higher Than Cold Condition

Non-heat treatable alloys may be heated to a temperature in the range of 300° or 400°F without any serious loss in mechanical properties and the most commonly used heat treatable alloys of 6061-T6 and 6063-T6 may be bent in this manner if the times and temperatures are observed:

300°	**325°**	**350°**	**375°**	**400°**	**425°**	**450°**
100–200 hr	50–100 hr	8–10 hr	1–2 hr	30 min	15 min	5 min

Charts are available for other heat treatable alloys. However, it is cautioned that any reheating sufficient to improve the formability may lower the resistance to corrosion to an undesirable degree, although loss in strength will not normally exceed 5%.

Loss in strength will not normally exceed 5%. However, remember that reheating sufficiently to improve the bendability may lower corrosion resistance to an unacceptable level.

Bending for Pneumatic Conveying

Both pipe and tube are used extensively for conveying purposes among a wide variety of industries, but principally for food and dry chemicals.

Although some carbon steel is used, stainless steel and aluminum are more common and are almost invariably thin wall since pressures are relatively low.

The sizes range from 1″ OD tubing up to 12″ IPS pipe—occasionally slightly larger—with some oval and rectangular material. Radii will vary from as little as 1.25D for air only to as much as 10D or even greater, particularly in the upper range of pipe sizes where radii of 96″, 120″, and 144″ are not uncommon. (Figures 3-2 and 3-3.)

Most of the bending for pneumatic conveying is done either in house by the companies that specialize in this field or by a very few job shops that have the size and type of machinery and tooling for the large radius requirements.

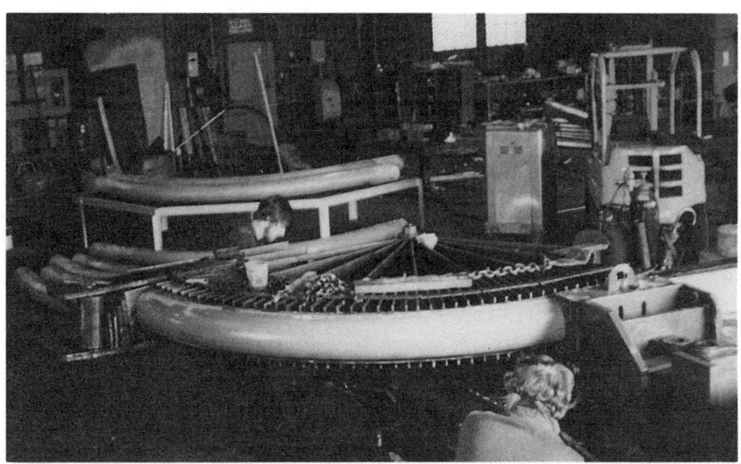

Figure 3-2. 6″ Schedule 20 stainless steel pipe being bent on a 72″ C.L.R. die. Note the clamp as being an integral part of the die. This home-built machine made in 1974 has made thousands of bends ranging in size from 2″–8″. (Courtesy of Tubebend Ltd.)

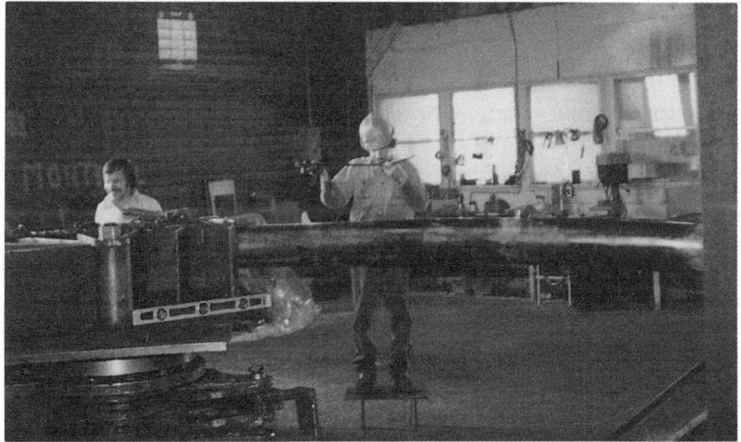

Figure 3-3. The same machine in Figure 3-2 being used to "bump" out an 8″ bend on a 10′ radius using a 72″ die. This outdated method is rapidly being replaced by induction bending. (Courtesy of Tubebend, Ltd.)

As one can imagine, a set of tooling for bending 12″ Schedule 20 or .250″ wall aluminum or stainless steel pipe on a radius of 120″ or 144″ is not a small investment, and a number of the machines used for this type of bending are of the "home built" variety, designed specifically for that purpose. These machines are relatively inexpensive to build, as are the dies, clamps, and pressure dies. One 8″ machine built in house in 1974 for something less than $20,000 is still in constant use today. An 8″ die for a 90° bend cost at that time about $2,500 to fabricate.

Today, of course, the same items, machine and die, would probably cost five to ten times these amounts, depending upon whether they were built in house or by a machinery manufacturer.

Unlike pressure piping where pipe ends are usually joined either by butt welding or by flanges (butt-weld or slip-on), which are welded to the pipe ends, most air-conveying pipe is joined by simple couplings: either Victaulic style couplings where one would have a roll groove in thin wall pipe or a cut groove to hold the clamps in Schedule 40 or heavier pipe, Morris couplings, or specially designed pressure couplings which are quite simple and easy to attach. The pipe or tube ends are square cut with no other special preparation required and butted up to each other. The coupling itself usually is slipped on, but may in some cases be of the wrap-around type, with neoprene, rubber, or asbestos gaskets. It is secured simply by tightening the bolts. See Figure 3-4.

Because conveying pipe and tube often carry abrasive material, a variety of treatments to the internal surface are available in order to reduce abrasion, both of the pipe or tube and conveyed material such as polyethylene pellets where abrasive action can cause fines and streamers.

Shot-blasting, hard-surfacing or ceramic lining of the pipe or tube are some of the options available. A number of processes are proprietary items. Treatment is performed only by those companies that specialize in the application and have the rights to it.

Not all applications are applicable to all materials, because the nature of the application process inhibits its use with certain materials. Ceramic lining, for example, is applied in powder form

Figure 3-4. Pneumatic conveying bends. Note the large sweeping radius and clamps for connecting pipe bends. (Courtesy of Allied Industries.)

at about 1500°F to a thickness of .008″ to .012″, but this high a temperature precludes application to aluminum tube or pipe, so the process is normally applied to carbon steel only.

Aluminum pipe and tube is used extensively for air conveying, but because of its very nature any hard surfacing process that involves even relatively low temperatures—about 450° or so—is precluded. Development of anodic hard-surfacing of aluminum bends is presently indicating successful results. Research is also being conducted into an entirely new method that will permit hard-surfacing treatment of any metal pipe or tube bend at low cost and in minimal time, without temperature or other material form (such as powder) implications. This particular process would be equally applicable to other conveying fields, such as slurry pipelines where abrasion is a constant problem.

Heat Exchanger Tube Bending

Boiler and heat exchanger tube bending, like pneumatic conveying, are specialty fields. Some companies may do both types or may concentrate on only one of the two.

Of the two, heat exchanger tube bending is by far the simpler since bends are almost without exception confined to small-diameter material with a maximum of 1″–1½″ OD and almost invariably 180° of bend. Lengths, of course, may vary considerably ranging anywhere from a few feet to 80 feet or even longer. In Europe lengths have been as much as 120 feet. Because the bend is almost invariably in the middle of the tube length, plenty of space is required since the swing area needed for a 60-foot tube will be at least a 30-foot radius covering 180° around. Space is also needed for storage of material prior to bending as well as for finished bends.

Heat exchangers may require return bends on 30 or 40 different radii, starting as low as 2–3D—sometimes even less than that—and going up to centerline diameters of 2–3 feet or more. The total number of tubes can run into hundreds or even thousands so ample storage and working space is absolutely essential.

There are two basic approaches to heat exchanger tube bending. One is to use dies exclusively—precision made with accompanying support precision tooling for the smallest radii (Figure 3-5) and wooden or aluminum dies for the larger radii. In some cases, conical dies may also be used.

Tight-radius bends may have to be done by the draw method, whereas the larger radii can usually be handled by compression bending.

The other method of bending developed by some shops does away with dies where noncritical radii are concerned, which is any radius that does not require precision tooling and by which the tube can be bent by the compression method.

In most cases, they have modified an older machine—a Pines #2 seems to be the most popular one used—by mounting a flat steel plate about 6 ft in diameter on top of the die plate and then converting the swing arm to function as a compression bend arm while the pressure die is converted to a clamping function.

Figure 3-5. Heat Exchanger Tube Bending. The ½″ OD × .049″ model was not allowed to be touched by hand; hence the white gloves. Note the amount of overbend required to compensate for springback, common in the type of bending.

Instead of using dies, a special type of roller chain is used slightly off-center to provide an equal center-to-center dimension as turns of the chain are added from a second (storage) plate attached to the machine.

To increase radius all one has to do is to unlock the plate and rotate it for as much of a turn or as many turns as necessary by adding more chain to get the desired center-to-center bend diameter. Then lock up the plate again, adjust the clamp and roller arms for the increase in radius and start bending again.

This is probably the least expensive method to make 180° heat exchanger tube bends, because no die is needed and changing a radius is very simple, quick and easy, thus cutting downtime to a minimum.

The benders who have this type of equipment guard it carefully. They rarely if ever allow photographs. However, a bit of ingenuity will always enable one to design something along these lines that will do the job quite effectively. (See Figure 3-6 for one of these rare photographs.)

Another useful modification is to add a support system with a cantilever swing arm to hold the straight material in front of the bend. This really is essential if one is to keep labor input at a minimum.

Imagine having an 80-ft length of 1/2″ OD by .049 tube to bend 180° with almost 40 ft extending out of the front of the machine and then swinging it 180°. It would take three or four assistants holding the 40-ft length with the one furthest from the bend having to run like mad every time a bend was made. Maybe in

Figure 3-6. A "modified" Pines #2.

China, with really cheap labor, it might be feasible, but not very practical. The cantilever supporting swing arm is the only real solution.

Of course, there are a few heat exchanger tube bending machines made by commercial manufacturers, but most of these require the use of dies. So far no manufacturer has seen fit to design and build a machine using the chain method.

For this particular type of bending, the homemade models or modified units seem to be the best and by far the most economical solution to date.

Boiler Tube Bending

This field is also considered to be a specialty area, although many job shops do have the capability of undertaking some form of boiler tube bending from time to time.

Broadly speaking, however, boiler tube bending is either relatively simple and straight forward, on fairly generous radii, or somewhat complex with a whole variety of compound bend requirements. There is also a good deal of repetition involved since quantities of tubes with the same radius and DOB are required for a single boiler.

For example, in a black-liquor recovery boiler, the front and rear walls may each have 60 or more tubes with a 90° bend at the bottom; thus the bending can be accomplished with a single setup. However, openings in the tube wall may require a considerable variety of different configuration bends calling for compound tooling and frequent changes of setup for varying DOB's and planes. This can be seen in Figure 3-7A and B.

A further complication is that the tube lengths can easily run 40 ft and longer with 18 or 20 compound bends in a single length. If these bends can be achieved satisfactorily without need for a mandrel it is entirely possible for all the bending to be done on the 40-ft length if the machine has hitch-feed capability with straight-through chucking.

Since a great deal of boiler tubing is relatively heavy-walled in the area of .109″ to .200″ in the 1½″ to 2″ OD size range, empty

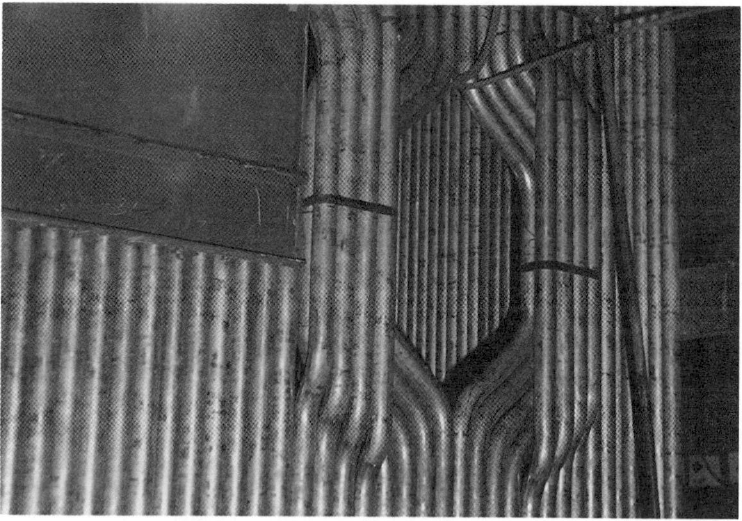

Figure 3-7. Typical tube bends (A) welded and (B) unwelded. (Courtesy of Canadian Industrial Tubebend.)

bending is usually achievable and can also be enhanced by an Incremental Pressure Die Assist System (IPDS) as outlined in Chapter 4 on Tooling.

Like most other forms of draw bending, if one has the right machinery, tooling and support equipment, bending is seldom problematical. Some bends, such as those on generous, noncritical radii, are best accomplished on a press bender for quantity production as illustrated in Figure 3-8. Membrane tube walls are usually bent by this method where wall sections of up to 10 ft in width can be bent in one single operation at considerable time and cost savings.

Certain support equipment is also necessary for boiler tube work, particularly swaging and/or upsetting machinery. Because

Figure 3-8. Membrane tube wall bender. (Courtesy of Schwarze-Wirtz A.G.)

boiler tubes are nested very closely together and usually have to be welded into a header, the ends must be reduced or swaged down in order to allow room for welding the tubes into position.

Several varieties of swaging machinery, mostly cold process in nature, can perform this function as well as some that hot form the tube ends. This latter method permits end forming of tube with extra heavy wall or large-diameter material that may be beyond cold forming capabilities in much the same sense that only a hot process can accomplish certain types of bending beyond the capabilities of cold machinery.

The most efficient method, of course, is by induction since this is far quicker than gas-fired or other methods for a wide variety of size ranges, wall thicknesses, and upset configurations.

One particularly difficult aspect of boiler tube bending is in forming 180° return bends for superheater tubes where the radius is as little as ½D. The normal approach to meeting this type of bend requirement is first to heat the tube in the bend area, then upset it to increase the wall thickness and the OD by pushing the tube on either side in toward the bend area.

Next the tube is cold bent 180° on a 2 to 3D radius and then heated, in the bend area only, of course. Then in several separate, subsequent stages it is compressed down. First, dies are used and then direct pressure application close to the tangent points squeezes the return bend down to the radius required before finally being pressed in a sizing die. The intrados of the bend area thickens very substantially but the extrados, having been upset from .200″ wall to say .300″ wall to start with, loses the extra thickening during the initial and subsequent bend stages and finishes close to its original thickness. The process is somewhat slow and highly labor intensive, but there does not appear to be any method better than this at present. With the introduction of small-induction bending machinery, however, there is now no doubt that development of a suitable machine and feed system for this particular aspect of boiler tube bending is in the offing and should be able to reduce a tedious, labor intensive operation to a one man situation.

4

Tooling and Lubricants

All pipe and tube bending, regardless of how it is achieved, requires tooling of some kind or another. Even if one end of a pipe is tied to a tree and the other end is pulled around mechanically, the tree serves as a primitive form of tooling. The tighter the radius and the thinner the material, the more tooling needed to achieve a satisfactory bend.

Contrary to some opinions, it is not the machine that makes the bend. It is the tooling. The machine merely supplies the power and contains and directs the forces involved.

Electricians mainly use a simple hand bender to turn corners in conduit piping. The bender is in fact the tooling while the electricians provide the force required to make the bend.

Tooling, therefore, is a very critical part of bending. Since the majority of all pipe or tube bending is by the rotary draw process, we will deal with that in detail and then outline and illustrate the tooling applicable to other methods.

Bending square and rectangular material is much more difficult than round tubing (Figure 4-2). Multiple ball mandrels and wiper dies are essential. To help determine what tooling is required, refer to the tooling selection charts calculating application for a round tube and add 20 to OD to wall factor.

Bend dies are usually split horizontally to permit the top half to be raised and lowered.

All rotary draw cold bending requires at least three items of tooling and may go to as many as six. (Figure 4-1 and 4-2). They are:

The Die. Sometimes this is referred to as the Center Former. This is what determines the radius of the bend and for each different radius there has to be a different die. In simple format, the tooling looks like this:

The radius is taken from the center point of the outside and the material is enclosed by the die to half the tube outside circumference. A straight section extends back in length usually at least

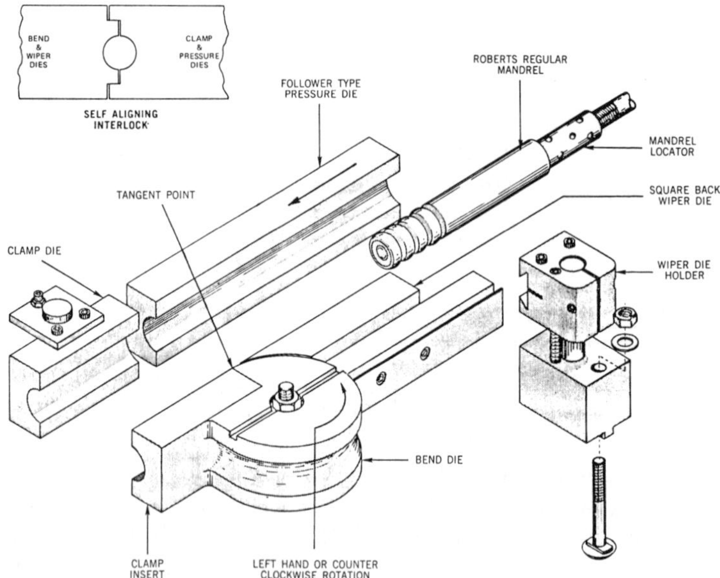

Figure 4-1. Typical tooling set (not including clamp arrangements or "booster" bending. (Courtesy of Tools for Bending, Inc.)

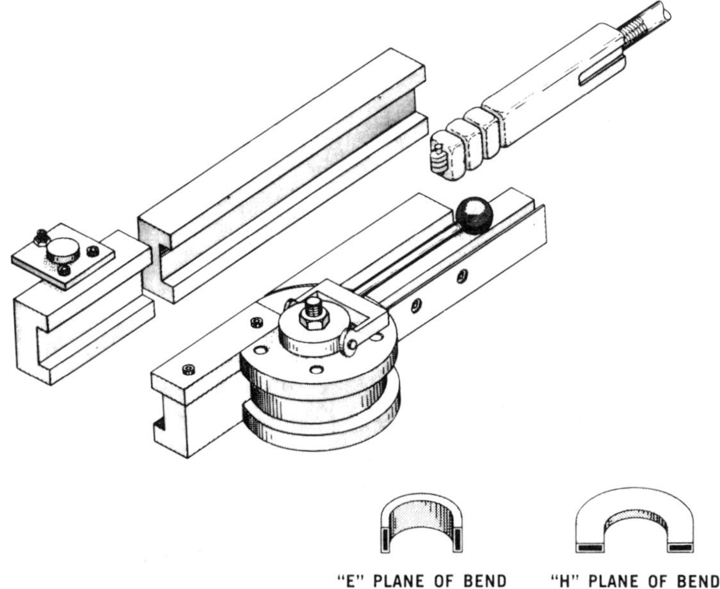

Figure 4-2. Typical tooling configuration for square or rectangular tubing. (Courtesy of Tools for Bending, Inc.)

twice the diameter of the material and this is called the Clamp Section, which is used to hold the pipe onto the die in such a manner that the bend will start at the tangent point, which as we explained earlier is the transition point where the bend starts or finishes.

A keyway on the underside of the die locates it correctly in position and it is kept firmly in place with a large nut or sometimes a wedge at the top of the center pin.

The Clamp. This is a matching half to the clamp section of the die and by means of a toggle or direct acting force holds the material firmly in place sufficiently to prevent the pipe from slipping through the two sections. If slippage should occur, inevitably, it will affect the bend quality and almost invariably result in a wrinkle forming on the inside of the bend itself.

Most draw bending machines are operated on the principle of a swingarm which provides the rotational track for the die and clamp. One U.S. manufacturer, however, has a different method which does away with the need for a swingarm and mounts the clamp directly onto the top of the die itself. In some respects, this is a cleaner, more efficient method for clamping but has certain drawbacks in regard to making compound bends.

The Pressure Die (static type). This is the third of the minimum key items of tooling. In its simplest form, it is very similar to a clamp although considerably longer and is referred to as a *Static Pressure Die.* It lies in the same plane as the clamp on the opposite side of the die and holds the material straight as it feeds into the die. Its prime function is to transfer the material at the tangent point directly to the die.

Because the pipe or tube is clamped tightly to the die, as the die rotates, the material is drawn along the pressure die and transfered to the die at the tangent point. It is necessary to use lubricant between the material and the pressure die to reduce friction and drag. Although good bends can be made by this method, it is not satisfactory for critical bending and is now rapidly going out of style.

Draw bending, using a static pressure die, tends to create too much stretch and therefore wall thinning beyond an acceptable level.

The Pressure Die (follower type). As a means of reducing stretch and wall thinning, the pressure die was extended and set onto a roller box or backed against a slide plate that allowed it to travel with the material, transferring it at the tangent point and moving on straight out of the way as the material stayed with the die. This advancement over the static pressure die was known as the follower type and made for better quality bends. There still remained, however, considerable stretch of the material and further means were sought to alleviate this problem.

The Pressure Die (assisted). To help contain the wall thinning, a hydraulic boost was attached to the follower die and by balancing pressure against rate of travel, it became possible to assist the material along at the same speed as it was being drawn by the rotating action of the die thus considerably reducing drag.

Even so, there was constant need to reduce wall thinning, so a further refinement was added to the pressure die.

The Pressure Die (boosted). In addition to having a hydraulic cylinder push the pressure die toward the tangent point at the same speed as it is being drawn, a clamping system was devised that either held the material tightly onto the pressure die itself at the rear end or separately from the pressure die, so that it could be pushed even harder than it was being drawn. This tends to compress the pipe toward the tangent point and thus contain the wall thinning (stretch) more effectively.

Most manufacturers clamp the pipe firmly to the pressure die and leave it at that, but one manufacturer went a step further and now has two boost cylinders which, between them, generate sufficient force to actually push the material out of the clamping area.

Careful adjustment of the various pressures involved now results in excellent control of wall thinning, but it is unlikely that mechanical measures will ever be able to reduce the degree of wall thinning much beyond the present level without introducing an element of heat.

The evolution of the application of the pressure die from static to follower to assisted to boosted and double boosted has been directed mainly toward making a 1½D pipe bend with wall thinning comparable to or better than that of a standard butt weld fitting, which under ASA B16.9 is required to be not more than 12½%.

So far, this has proven to be an almost unattainable objective except when bending A 587 carbon steel pipe and some stainless steels.

Without a suitable means of compressing the extrados and reducing the compression of the intrados of the pipe bend, it seems unlikely that wall thinning control will advance much further than it is right now, at least insofar as cold bending is concerned. There are two more items of tooling which are essential to success in bending thin wall material on tight radii. It is possible with the tooling described so far to bend heavy wall pipe and tube on quite critical radii, particularly tube which is generally more ductile and bendable than pipe, and it is possible to get down to as little as 2D and sometimes even 1D with the tooling just described.

Generally speaking, however, the tighter the radius and the thinner the wall, the more need there is for an internal mandrel and an external wiper die. See Figures 4-3, 4-4, 4-6, and 4-7.

The Mandrel. There are several different types of mandrel which are basically designed for internal application to the pipe or tube in order to assist in maintaining the shape of the pipe after it is transferred from the pressure die to the die.

In its simplest form, it is a solid piece of steel or other suitable material fitting closely inside the pipe or tube yet not so close

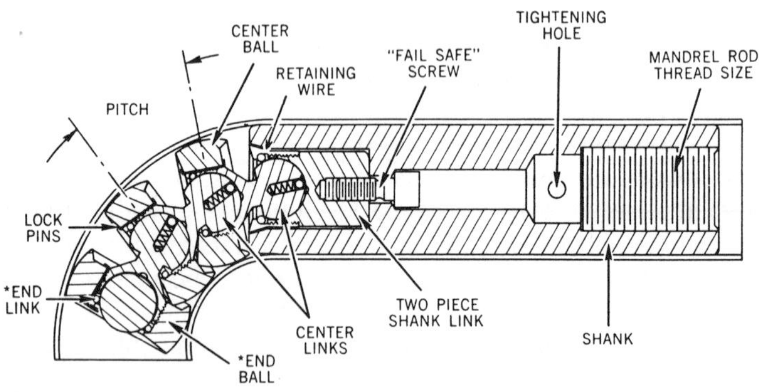

Figure 4-3. Schematic view of a Roberts regular 3-ball mandrel showing linkage of balls allowing for 360° articulation. (Courtesy of Tools for Bending, Inc.)

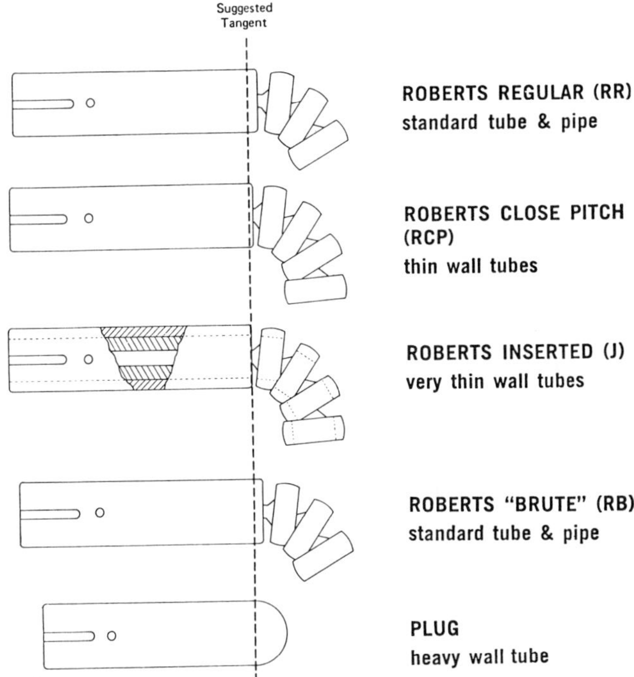

Figure 4-4. Various types of mandrels. Note that the ultra thin wall bending the nose is on the tangent line and beyond this for all other wall thicknesses. (Courtesy of Tools for Bending, Inc.)

that it can jam. Generally the clearance is in the region of .012″ all around and the nose of the mandrel is positioned so that it projects slightly past the tangent point of the bend. The nose works against the outside wall and helps to keep it pushed out on the outside radius of the bend. Without a mandrel, the wall would tend to move inward toward the neutral axis and thus flatten out.

This tendency increases as the radius of the bend is decreased and as thinner wall material is used. To counteract against this, articulated balls are added, very slightly less in diameter than the shank (main body) of the mandrel. These balls help to maintain the shape of the material after it passes over the nose of the mandrel. Figure 4-3 shows a diagrammatic view of a Roberts Regular three-ball Mandrel showing how the linkage of the balls allows for full 360° articulation, and how the balls function against the inner and outer walls of the tube to maintain ovality.

For ultrathin material, there may be as many balls as the developed length of the bend, but because they are articulated, when the bend has been completed and the mandrel is retracted in order to remove the workpiece, they can be pulled back through the body of the bend and into the straight section before removing the tube or moving the pipe forward for the next bend. Note in Figure 4-4 that for ultrathin wall bending the nose is *on* the tangent line and for all others projects beyond. See also Figure 4-5.

Since the mandrel and balls are actually working against the surface of the inside of the pipe or tube it is necessary to use lu-

Figure 4-5. A 2-ball mandrel for stainless steel bending. Note the articulation and the lubrication groove for 360° dispersal of lubricant. (Courtesy of International Piping Systems, Ltd. and Schwarze-Wirtz A.G.)

bricants in order to reduce friction. Most mandrels today are designed with lubrication holes. The lubricant is dispersed either manually or automatically while the tube is being loaded onto the machine as well as during the initial stage of the bend and even for the entire duration of the bend. The fit of a mandrel is critical and may often present problems of its own, particularly when it applies to pipe.

Too small a mandrel means a loose fit and thus loss of effective control; too large and the mandrel can jam in the pipe. Generally speaking, as stated previously, there needs to be a clearance of about .012″ all around, but with some material having a ± 10% tolerance on the ID that can be a real problem and affect the quality of the bend from one to the next.

Tube is usually much more consistent in terms of ± tolerances and does not create as many problems as pipe does. Ideally, of course, constant wall thickness would be the answer, but that is rarely possible, particularly with pipe. So one generally finds tolerances of "not to exceed . . ." when prior experience has shown what is readily achievable.

The Wiper Die (Figure 4-6). This particular item of tooling is designed to assist in *preventing* wrinkles from forming in the intrados of the bend rather than wiping them out. Once a wrinkle has occurred it cannot be removed, at least during the bending process, and is virtually impossible to correct after. Thus the wiper, so named because it *wipes* against the die, is a preventive, as opposed to corrective, piece of tooling. See Figure 4-7.

It is very precisely tooled to match the radius of the die as well as the pipe size. The tip or feather is razor sharp when new and must always be protected when not in use. Since considerable pressures are applied to the wiper and the material travels along its length during the bending process, there is a great deal of erosion at the tip. Consequently, the wiper must be reworked fairly frequently, particularly when used with coarse surface material which accelerates the erosion problem. It is wise, therefore, to have at least one spare wiper if one is going to do volume work of this nature.

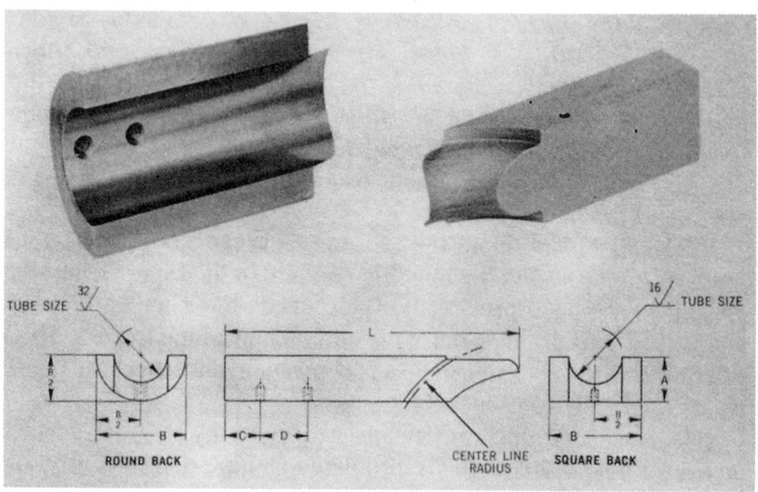

Figure 4-6. Typical configuration of a wiper die. (Courtesy of Tools for Bending, Inc.)

Figure 4-7. Rear view of a wiper die showing how it fits into the die and "wipes" against it. (Courtesy of International Piping Systems, Ltd. and Schwarze-Wirtz A.G.)

Similarly, because the wiper is in abrasive contact with both the die and the bending material, lubrication is essential for both areas in order to reduce the abrasive effect.

Positioning of the wiper is as critical as it is for the mandrel. Generally speaking it is raked back from 1–3 degrees away from the pipe to eliminate drag. To determine whether a wiper die is necessary, always refer to the tooling charts (Figure 4-8). In

Ratio of $\frac{\text{centerline radius}}{\text{tube outside diameter}}$

Ratio of $\frac{\text{tube outside diameter}}{\text{wall thickness}}$

Bend Angle		1 XD 90°	1 XD 180°	1.50 XD 90°	1.50 XD 180°	2 XD 90°	2 XD 180°	2.50 XD 90°	2.50 XD 180°	3 XD 90°	3 XD 180°	5 XD 90°	5 XD 180°
10	Ferrous	P	P	P	P	P	P	P	P	-0-	P	-0-	-0-
	Non-Ferrous	P	P	P	P	P	P	P	P	P	P	-0-	-0-
20	Ferrous	RR-1	RR-1	RR-1	RR-1	RR-1	RR-1	P	P	P	P	P	P
	Non-Ferrous	RR-1	RR-2	RR-1	RR-2	RR-1	RR-2	RR-1	RR-2	RR-1	RR-2	P	P
30	Ferrous	RR-2	RR-3	RR-2	RR-3	RR-2	RR-3	RR-1	RR-2	RR-1	RR-2	P	P
	Non-Ferrous	RR-3	RR-3	RR-3	RR-3	RR-3	RR-3	RR-3	RR-3	RR-2	RR-3	RR-1	RR-2
40	Ferrous	RR-3	RR-3	RR-3	RR-3	RR-3	RR-3	RR-3	RR-3	RR-2	RR-3	P	RR-1
	Non-Ferrous	RCP-3	RCP-4	RCP-3	RCP-4	RR-3	RR-4	RR-3	RR-3	RR-3	RR-3	RR-2	RR-2
50	Ferrous	RCP-3	RCP-4	RCP-3	RCP-4	RR-3	RR-4	RR-3	RR-4	RR-3	RR-3	RR-1	RR-2
	Non-Ferrous	RCP-4	RCP-5	RCP-4	RCP-5	RCP-4	RCP-5	RCP-3	RCP-4	RR-3	RR-4	RR-2	RR-3
60	Ferrous	RCP-4	RCP-5	RCP-4	RCP-5	RCP-3	RCP-4	RR-3	RR-4	RR-3	RR-4	RR-2	RR-3
	Non-Ferrous	RCP-4	RCP-5	RCP-4	RCP-5	RCP-4	RCP-5	RCP-4	RCP-5	RCP-3	RCP-4	RR-3	RR-3
70	Ferrous	RCP-4	RCP-5	RCP-4	RCP-5	RCP-3	RCP-4	RCP-3	RCP-4	RR-3	RR-4	RR-3	RR-3
	Non-Ferrous	J-4	J-5	J-4	J-5	RCP-4	RCP-5	RCP-4	RCP-5	RCP-4	RCP-5	RR-4	RR-5
80	Ferrous	RCP-4	RCP-5	RCP-4	RCP-5	RCP-3	RCP-4	RCP-3	RCP-4	RCP-3	RCP-4	RR-3	RR-4
	Non-Ferrous	J-4	J-5	J-4	J-5	J-4	J-5	RCP-4	RCP-5	RCP-4	RCP-5	RCP-3	RCP-4
90	Ferrous	J-4	J-5	J-4	J-5	RCP-4	RCP-5	RCP-4	RCP-5	RCP-4	RCP-5	RR-4	RR-5
	Non-Ferrous	J-4	J-5	J-4	J-5	J-4	J-5	J-4	J-5	RCP-4	RCP-5	RCP-4	RCP-5
100	Ferrous	J-4	J-5	J-4	J-5	J-4	J-5	RCP-4	RCP-5	RCP-4	RCP-5	RCP-4	RCP-5
	Non-Ferrous	J-5	J-6	J-4	J-5	J-4	J-5	J-4	J-5	J-4	J-5	RCP-4	RCP-5
125	Ferrous	J-4	J-5	J-4	J-5	J-4	J-5	J-4	J-5	RCP-4	RCP-5	RCP-4	RCP-5
	Non-Ferrous	J-5		J-5	J-6	J-4	J-5	J-4	J-5	J-4	J-5	RCP-4	RCP-5
150	Ferrous	J-4	J-5	J-4	J-5	J-4	J-5	J-4	J-5	J-4	J-5	RCP-4	RCP-5
	Non-Ferrous			J-5		J-5	J-6	J-4	J-5	J-5	J-6	J-5	J-6
175	Ferrous	J-5		J-5	J-6	J-4	J-5	J-4	J-5	J-4	J-5	RCP-5	RCP-6
	Non-Ferrous					J-7		J-7	J-8	J-6	J-7	J-6	J-7
200	Ferrous			J-6		J-5	J-6	J-5	J-6	J-4	J-5	J-6	J-7
	Non-Ferrous							J-9		J-9		J-9	J-10

KEY

P — Plug
RR — Roberts Regular
RCP — Roberts Close Pitch
J — Roberts Inserted

NOTE *A wiper die is suggested for applications falling beneath the dotted line

Number indicates suggested number of balls

Roberts "Brute" mandrels super strength and high production available in RR and RCP configurations.

Figure 4-8. Typical tooling chart.

some cases, doubt may exist as to whether a wiper is really necessary and in these instances one can try a test bend and should wrinkles appear then the answer is use a wiper. If there are no wrinkles, then the wiper is probably not necessary. See Figures 4-9 and 4-10.

In order to better control the three key factors that govern the quality of a bend (wall thinning of the extrados, compression of the intrados, and distortion, which includes loss of ovality), some of the tooling companies and machinery manufacturers started to research the tooling required to provide more effective

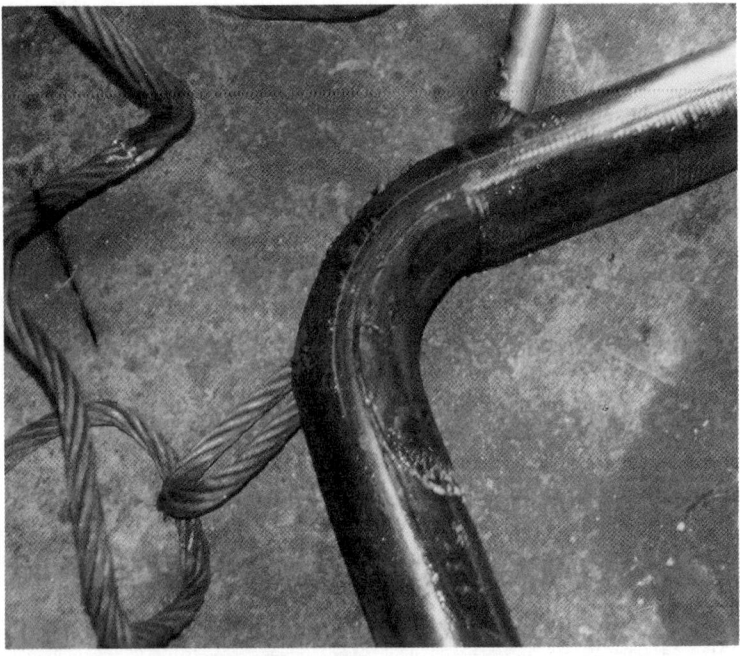

Figure 4-9. Poor-quality bend. In this case, the mandrel was retracting prematurely, producing a flattened bend. Note the wiper die mark on the intrados of the bend. (Courtesy of International Piping Systems, Ltd. and Schwarze-Wirtz A.G.)

Figure 4-10. A broken bend resulting from improper or incorrect tooling set-up or machine malfunction. (Courtesy of International Piping Systems, Ltd. and Schwarze-Wirtz A.G.)

control of these three factors. Among them Tools for Bending, Inc. (TFB) conducted extensive research and developed what they called an Incremental Pressure Die Assist System (IPDA System) that took into account the need for varied speeds and pressures during the progress of a bend.

Because material "flows" once it reaches the yield point and tends to do so both at the start of the bend where it precedes the front tangent line and at the end of the bend where it moves behind the rear tangent line, it is necessary to reduce the pressure in these two areas in order to minimize the tendency to create a bulge (Figure 4-11).

Research by TFB using mild steel, aluminum and stainless steel tubing with wall factors as low as eight and as high as 71 for 180° return bends resulted in determining the need for reduced pressures and speed for the first 21° of bend, increased for the next 160° and reduced again for the final 8° in order to achieve optimum results. While there was still some flow beyond the tangent points at either end of the bend, it was found to be minimal.

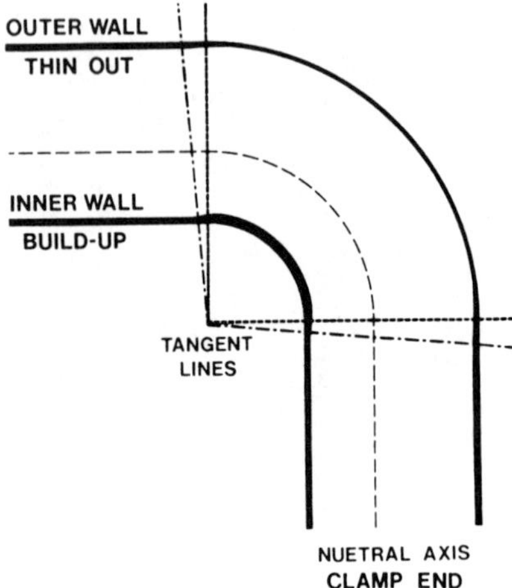

Figure 4-11. Wall thinning and thickening actually start beyond the tangent lines. (Courtesy of Tools for Bending, Inc.)

Ovality reduction using the IPDA method in comparison to constant speed and force improved substantially as did wall thinning. However, even better results were achieved with use of an oscillating mandrel rather than one in a fixed position.

Although a number of bending machine manufacturers now incorporate variable speed and pressure control in their newer machines, there are still a great many older ones in use that do not have this degree of sophistication. They can in many cases now be retrofitted with the IPDA System.

Empty Bending

A further advantage to using an IPDA System comes with the development of "heart-shaped" tooling for empty bending. When the wall thickness of a particular tube is sufficient for it to be bent without the need of a mandrel, the savings in tooling and time costs can be quite substantial. However, further cost effectiveness can be achieved by using thinner wall tubing, but under normal circumstances this would necessitate the use of mandrel tooling again.

Heart-shaped tooling (Figure 4-12) where the die and pressure die each have slightly different ID configurations and functions—the bend die inducing compression and the pressure die minimizing collapse and wall thinning reduction—permit bending of tubes with thinner walls than would normally be the case using standard tooling. Used in conjunction with an IPDA System, quite severe bends can be achieved at considerable cost savings with excellent quality results (Figure 4-13).

Tooling technology is a very important element in bending both from a cost and product quality aspect. It takes longer to load a tube over a mandrel than it does without one. The thicker the wall a tube has the more it costs. If one can eliminate the

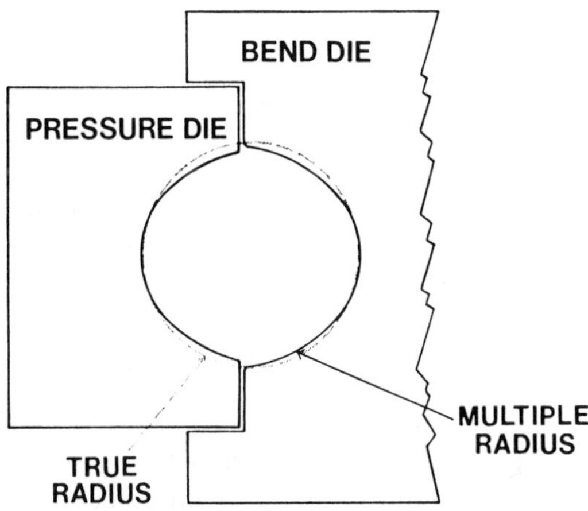

Figure 4-12. Actual tube OD (dotted line) in relation to the pressure die and bend die ID. (Courtesy of Tools for Bending, Inc.)

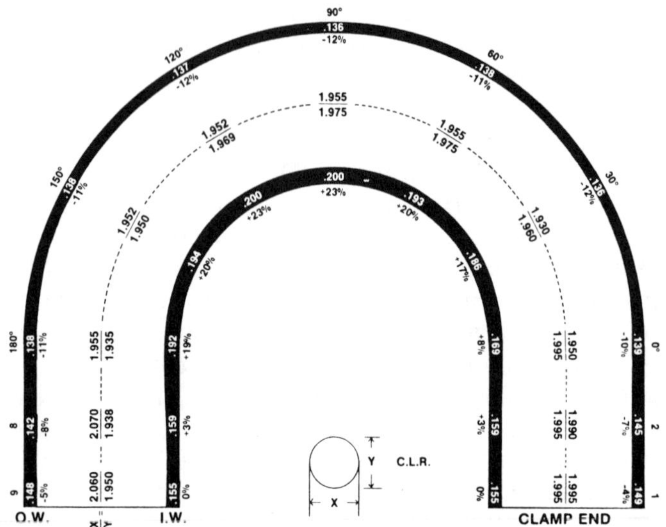

Figure 4-13. An example of wall thinning and wall compression results in 2″ OD × .155″ tube bent 180° on a 1½ D (3″) CLR using Incremental Pressure Die Assist in conjunction with "Empty Bend" tooling. (Courtesy of Tools for Bending, Inc.)

mandrel and wiper die not only have these costs been eliminated, but production time has been increased and over-all cost decreased. Using thinner wall material again results in lower cost, and when one adds the factor of product quality improvement, all arguments against capital expenditure for the necessary equipment to achieve these results become moot.

Lubricants

The use of suitable lubricants is critical in nearly all the aspects of radial draw bending, other than "empty" bending where no mandrel is required. Lubricants are seldom necessary for most other methods of pipe or tube bending such as compression, ram, coil, hot slab or induction. But the mere fact that a mandrel has to be used and a wiper die, if necessary, dictates the use of proper forming lubricants. In the early days of draw bending every conceivable type and kind of grease, oil, soap, or handy substance was used. Even Ivory Soap™ flakes have been known to be used as a bending lubricant and many shops saved what was considered unnecessary expense by resorting to old hydraulic fluid as a bending lubricant.

However, as technology improved, radii became tighter, wall thickness became thinner, and bending more critical, special compounds and lubricants were developed. Many of these were petroleum-based and often created problems of their own. For one thing, removal after bending was tedious and bends had to be cleaned out in or with special solvents. Some materials could not accept the effect petroleum-based lubricants had on the pipe or tube material itself and, more recently, there was a matter of pollution caused by the disposal of waste liquids from the cleaning process.

Today synthetic nonpetroleum-based lubricants are used extensively because of their many advantages.

Designed primarily to promote lubrication, reduce corrosion and provide cooling, the synthetics have two distinct advantages—they are easy to clean out and, being biodegradable, are innocuous to health. Another advantage is that they are usually supplied in paste form and can be diluted down to whatever consistency may be required for the bending application. Dilution may be with water or mineral oil or other suitable nonpetroleum-based liquid compatible with the material to be bent.

The heavy consistency of these lubricants minimizes the widespread dispersal characteristic of oil-based lubricants and makes it much easier to keep the machine and its immediate working area cleaner. Shops which use oil-based lubricants are usually messy, to say the least, with "slurp" or other absorbent materials spread everywhere. Those using synthetic lubricants are far easier to keep clean as excessive lubricant can be wiped up quickly and does not spread in the same manner as oil does.

One of the few disadvantages is that tooling must always be protected after use since normally protective greases and oils have been removed. Some form of rust preventative should always be applied when tooling is stored for any length of time. Care must be exercised to clean out pocket areas, such as between the mandrel balls, where not only has the water-based lubricant collected, but it has almost certainly accumulated dirt, metal fines, etc., all of which must be removed prior to storage.

It is also important to determine possible interaction between synthetic lubricant properties and the bending material surface itself, to insure that it does not react in an adverse manner which could result in corrosion or staining.

The importance of using proper lubricants cannot be over-emphasized. Not just for mandrel application, but also for the wiper where it contacts the pipe surface and for the surface area between the wiper and the bend die. It is not unusual to find that three separate types of lubricant may well be needed, depending upon the nature of the material to be bent, the nature of the tooling itself, and the degree of severity of bend.

Thus, it is important to develop and maintain a close relationship with the supplier so that, at all times, one uses the correct lubricants and the correct consistencies for each application and that proper pre- and post-bend maintenance procedures are also followed. The result should be a relatively trouble free operation—at least as far as lubricants are concerned.

5

Operator Training and Equipment Maintenance

There are three key elements to bending—the machine, the tooling, and the operator. Generally speaking, the first two items are known factors. If one uses the right machine with the proper tooling one will be able to produce bends. How good or bad these will be, however, will be largely dependent upon the operator and how good or bad he himself is.

It is imperative, therefore, that operators be properly trained in the use of their machines, how to operate them correctly and (most importantly) safely, how to maintain them, how to apply the tooling and adjust it as needed and, finally, how to read a bend to analyze the results of their own handiwork.

To this end, training is vitally important. It is not too difficult to show a man a machine that has been set up for him, run through the bending cycle and operating procedures, and then set him to work making bends. However, sooner or later something will go wrong because you don't have a bender; all you have is a man who only knows how to push buttons or pull levers.

It takes time, effort, and a great deal of patience to train a bender properly and one must bear in mind that bending can be a very tiring and tedious job. After all, a bender is on his feet all day long often having to use a lot of muscle power loading and unloading pipe. Handling small OD, light gauge tubing is one thing, but working with 4″ and larger pipe can be very demanding physically and this hampers mental concentration.

Mistakes will usually occur late in the morning or afternoon when the man is tired and not concentrating properly. In a mixed job shop, he can look around and see welders sitting at their job or machinists reading a magazine while a lathe or milling machine does all the work. Benders have to be properly motivated and they must not only have an inquiring mind but a strong desire to do their job well.

Selection of personnel, therefore, is highly important. If one can identify the right type of individual and then train him properly so that he learns and has the desire to go on learning (in bending there is always something new to learn) one will eventually wind up with a real craftsman.

Unfortunately this type of individual is a rarity and many shops have poorly motivated and poorly trained benders. The results will always show in the end product.

Training, therefore, must be planned and followed through carefully and methodically and should not be restricted to potential operators only.

Supervisory personnel need to be familiar with procedures as do sales and management. These people do not have to become skilled operators, but they do need to know what bending is all about, particularly the kind of bending their own company is involved with.

Some years ago we ran a bending training program in our shop in Canada, where the federal government paid 60% of the trainees' wages. The program ran for 26 weeks and at the end of that time, we had a few really good operators who knew what they were doing, and why, and were familiar with all seven machines that we had running in the shop.

If at any time a man was sick or unable to come to work—and that sort of situation invariably occurs during a rush job—we always had other operators we could pull off a less critical job and put them on the priority one without having to worry whether they could handle it.

Training in depth, therefore, is important. One man for one machine is fine if he is particularly good on that machine. But if you don't have a backup or two in case that man gets sick then you will have a problem.

The accompanying program outline is for draw bending. Each machine should have a program designed around its function and range of capabilities and a full scope program should include operating procedures for every machine in the shop, not just one. Machines have idiosyncrasies and no two identical models will ever be exactly the same.

Classroom instruction, both from a theoretical viewpoint as well as a place for periodic testing, is very important. The classroom atmosphere is usually quieter and cleaner than the shop and more conducive to assimilation of instruction. The training program outlined here is very basic and parts of it will be redundant where really modern machinery is concerned. It is shown purely as a guide and anyone setting up a training program must tailor it to serve their own needs.

Without a sound training program, one will not have slow benders, nor will one have fast benders—one will invariably have "half-fast" benders!

Sample Training Program
Cold-Draw Bending

Attendance:

Operators and Assistants (A)
Immediate Shop Supervisors (B)
Selected Management and Sales Personnel (C)

Introductory Theory and Basic Bending Procedures

Theoretical—Classroom (Students A, B, and C)

1. Introduction
2. Types of bending

3. Principles of draw bending
4. Terminology
5. Tooling
6. Lubrication

Practical—Shop

1. Introduction to the machine
2. Identification and designation of systems, elements and components
3. Safety—standard operating procedures
4. Maintenance—routine daily inspection procedures
5. Setup—tooling selection and mounting
6. Tooling setting and adjustments
7. Lubrication
8. Loading

Theoretical—Classroom: Students A and B only, C optional (Single bend and 2 bend spools)

Job Preparation

1. Material: type, size, schedule
2. Radius of bend, tangent lengths
3. Number of bends
4. Distance between bends
5. Plane(s) of bend(s)—single, 90° and 180°

Practical—Shop: Students A and B, C optional

Machine Preparation and Basic Bending

1. Inspection—routine maintenance procedure
2. Tooling selection and mounting
3. Machine and tooling settings—dry run
4. Machine and tooling settings—test piece
5. Workpiece loading

6. DOB, POB, DBB settings
7. First bend
8. Bend analysis
9. Setting readjustment clamp, pressure die, boost, mandrel, wiper, and DOB for springback
10. Second bend
11. Bend analysis
12. Workpiece removal (repeat as required for each student)

Quality Assurance and Dimensional Accuracy

1. Floor layout—3-4-5 method
2. Tangent marking
3. Manual, on-machine checking; use of set-square and Racine Miracle Point center finder
4. Floor check (repeat as required for each student)

Practical—Shop: Students A. B and C optional

1. Tooling changes
   - Mandrel only (schedule changes)
   - Die only (radius change)
   - Full change (pipe size change)
2. Tooling removal, cleaning and storage
3. Mandrel rod removal and cleaning
4. Machine inspection and shutdown (repeat as required for each student)

Review—All theoretical (Classroom)

Review—All practical (Shop)

Written test—(T & F and multiple choice)

Practical test—From drawing of workpiece to machine shutdown. Complete run through by each team of students either in pairs or groups of three.

Spool Production—Standard and heavy wall—multiple bends

Theoretical—Classroom: Students A, B, and C

1. Review Introductory theory and basic bending procedures.
2. Introduction to wall and bend factor calculations.
3. Developed length calculations.
4. Determination of multiple COP implications
5. Determination of clamp, tangent (DBB), and pressure die length factors.

Practical—Shop—Heavy wall-empty bending: Students A, B. C optional

1. Routine inspection
2. Setup and tooling mounting and adjustments
3. Determining first tangent point and DOB, DBB, and POB settings for all bends.
4. Unloading
5. QA and dimensional verification
6. Review of possible required adjustments
   Repeat for each group of students until all are practiced and familiar with procedures.
7. Tooling removal, inspection and shutdown.

Theoretical—Classroom: Students A, B, and C

1. Review previous theoretical section
2. Thin wall tooling—mandrel and wiper
3. Lubrication

Practical—Shop—Thin wall mandrel bending, single bends: Students A and B, C optional

1. Routine inspection
2. Setup and tooling mounting and adjustments, and pressure adjustments
3. Mandrel and wiper adjustments

4. DOB, settings
5. Bend evaluation
6. Mandrel and/or wiper readjustment
7. Bend evaluation, QA, dimensional verifications. Repeat for each group of students until all are practiced and familiar with procedures.

Practical—Shop—Thin wall mandrel bending, multiple bends: Students A. B, C optional

1. Repeat of thin wall mandrel bending using multiple bend COP settings.

Practical test (Students A. B, C optional)

Use other material, different work drawing

Drawing Interpretation and Bending Requirement Implementation

Theoretical—Classroom: Students A and B. C optional

1. Review basic procedures and spool production
2. How to read a drawing
3. Job requirement preparation
4. *Test project*

Practical—Shop

Test Project—Students A only. B, C optional

Students will be required to identify the correct material, have it cut to length, set up the machine, tooling, etc. and bend the workpiece after being given a spool or isometric drawing and no further information or assistance.

Students will be expected to be able to follow all procedures correctly and to produce a spool to the required configuration and QA standards.

Training—Engineering and Sales: Classroom

1. Introduction
2. Types of bending
3. Terminology
4. Shop machine, size, limitations, and bending parameters
5. Designing for bending
6. D factors—nominal and true
7. Material schedule implications
8. Bending factors, short tangents, compound bends, COP "kickdown" factors
9. Material take-off
10. Production time and cost estimating
11. R & D
12. Developing technology
13. CAD/CAM possibilities

Equipment Maintenance

One other aspect to pipe and tube bending that also has a strong bearing on end results is maintenance, probably the most neglected operation in many plants, yet it is the one that can, if it is performed correctly and regularly, help keep operating costs at a minimum and productivity and quality at a consistently high level.

As long as a machine keeps running little or no attention is paid to it, particularly if the shop is busy and has a lengthy backlog. Only when it breaks down and halts production is anything ever done and then more often than not, due to supervisory pressures, it is only a hasty patchwork repair job.

It is not uncommon to find shops where machines have been in use for a dozen years without ever having had regular scheduled maintenance. Temporary repairs have been exactly that—temporary—until the machine breaks down again and is fixed up enough to keep it running.

What usually happens with neglected machines is that the moving parts, i.e. clamp and pressure die slides, begin to wear and rather than have them repaired or replaced pressures are increased to compensate for slop in tolerances or settings. This continues until the operator gets to where he simply cannot get a good bend from his machine. At this point, the whole machine will probably need reworking. Apart from the considerable cost involved there is the likely loss of several weeks' production time.

The normal wear and tear on a machine is often compounded by lazy operators and shop supervisors who, in order to get the work out more quickly, will weld something that should be bolted. The best example of this is on older machines using a static pressure die. Instead of bolting the die into place, which may take some adjusting, the pressure die is positioned against the pipe and then welded quickly into position; no adjustment necessary. This happens time and time again until the back-plate and the die are completely shot and must be replaced. The final cost inevitably comes out far higher than the cost of the extra few minutes to mount the tooling properly.

Among some of the older benders there is a saying: "If it moves—weld it. If it doesn't move, take a hammer to it." Both are reprehensible when applied to a bending machine and the best precaution against this attitude is to bar any welding equipment or hammers whatsoever from even being near a bending machine.

Hammering a die down to seat it on the key-way is another common practice which, sooner or later, will cause problems with the key-way. Also, as is often the case, the idiot using the hammer misses and dents the lip of the die. This results in expensive rework of the die or one will get marked bends.

Maintenance should be planned, carried out, and checked on a regular basis from the first day a machine is put into operation. The term "maintenance" should be clearly understood by everyone in the shop. A great many operators and shop personnel interpret that word as meaning "repair", which is wrong. Perhaps "preventive maintenance" would be the best term, since the objective of any good program is to prevent the machine from breaking down by constantly maintaining it in proper operating condition.

A sound preventive maintenance program should have clearly defined objectives:

1. Minimize equipment failure.
2. Maintain proper tolerances.
3. Avoid unscheduled downtime.
4. Reduce hidden costs.

Machinery of any kind will fail at some time or another, but good preventive maintenance practices will help to minimize this possibility. A hydraulic hose that is rubbing will eventually wear through, so checking all hoses regularly and replacing any that show signs of wear will prevent one from blowing during production and causing a shutdown, not to mention the time and effort required to clean up the mess that follows.

Replacing wear strips as soon as they need replacement is far better than making adjustments to compensate for slop in tolerances.

Unscheduled downtime is always costly. The odds are 3:1 that equipment will fail during the afternoon or midnight shift or over a weekend and almost certainly in the middle of a rush job, resulting in overtime rates and excessive repair costs, not to mention possible late delivery penalties and irate customers.

Hidden costs, too, can run high. A vital part may be out of stock when needed urgently and have to be flown in from elsewhere. Instead of a no-cost local delivery we how have an airfreight shipment and the time and money spent just to pick it up at the airport and maybe, if it has come from overseas, the extra cost for customs clearance after normal working hours. You can be absolutely sure that Murphy's Law will prevail every time.

To achieve the desired objectives, it is vital to plan ahead. Set up a maintenance log, report forms, and schedule. Apply the program vigorously.

Indoctrinate, train and motivate all personnel involved, especially operators at the lower level and management at the upper level of the organization.

Schedule minor repairs and replacements for short, nonoperating periods such as evenings or weekends and major repairs for organized shutdowns, vacation periods, or holiday weekends.

Order and get parts well ahead of planned downtime and keep critical short wear-time parts and components on hand at all times.

The main thing is to correct a developing problem before it has matured into a crisis. This can only be done by regular inspection and the ability to recognize potential trouble before it becomes a reality. Daily inspection of the machine, components, tooling, etc. both immediately prior to starting daily production as well as last thing after shutting down for the day will go a long way toward catching a potential problem. Never leave an incipient situation for later attention and always report, in writing on the appropriate form, every little detail that needs attention.

U.S. airlines have the best safety record in the world because strict attention is paid to preventive procedures. If all our shops paid as much attention to this aspect of production, it would save untold millions of dollars and greatly increase our productivity.

So, in addition to the machine, the tooling, and the operator as key elements in bending, we should never neglect the fourth and most neglected one—preventive maintenance.

6

Support Equipment

In every bending shop there is support and ancillary equipment and machinery. Simply having a bender, whether manual, automatic, or CNC, is not enough to complete many bending jobs today.

Material may have to be moved, it may have to be cut, it may have to be welded, or it may require any one of many different kinds of further processing such as swaging, notching, Van Stone flanging, collaring, beveling, victaulic grooving, expanding, or drilling.

Thus, each shop will have a variety of equipment to take care of these requirements. Also, one doesn't just bend and process pipe. The pipe or tube may have to be cleaned before bending, washed out after bending to remove lubricants, painted, heat treated, stress relieved, or even galvanized. Then the finished item must be examined and probably tested to make sure it meets the appropriate standards, so there is a need for nondestructive testing equipment such as die penetrants, radiography, and ultrasonic or other forms of measurement.

The first and foremost item is usually material handling and movement of equipment. In many shops this is done by forklift and either overhead or boom cranes or a combination of all three. However, the most modern, up-to-date method is by automated material storage, movement, and processing, the best known of which is the Oxytechnik System, which originates from an engineering company of that name in West Germany. See Figure 6-1.

Basically the system utilizes a form of silo storage with pipe lying horizontally but stacked in tiers vertically according to size. A moving elevator system draws a length of pipe from the stack and lowers it to the conveying level where it is transferred to the line and moved forward to the first stage of processing. See Figure 6-2.

In most instances this would be a cleaning process, either internally or externally or both. For the external process the pipe is shot-blasted as it travels through the cabinet (Figure 6-3). Smaller diameters can be cleaned several at a time depending upon the capacity of the unit. The largest pipe sizes would, of course, go through singly.

For internal cleaning, a lance is used that permits only one length of pipe to be processed at a time.

After the pipe is cleaned it is then conveyed to the next stage, which may be any one of several actions. For shipyards, using no tighter than 2D bends, the pipe may be painted since the pipe can be bent in a painted condition. This is done in Avondale Shipyards in New Orleans. In 1981 they became the first company in the United States to install an Oxytechnik System.

Generally, the second stage in the system would be measuring and cutting. The pipe is measured for the exact length needed, taking into consideration the stretch factor for cold bending which reduces the cut length slightly or the compression factor for induction bending which, of course, increases the amount of material required. Both of these factors are pre-programmed into the computer beforehand since they are known data provided by the bending manufacturers—or should be.

Next, the pipe is square cut—usually by saw and more often than not *without* cutting oil.

Now that the workpiece is of the correct length either for further fabrication or bending or both, it is again conveyed by the

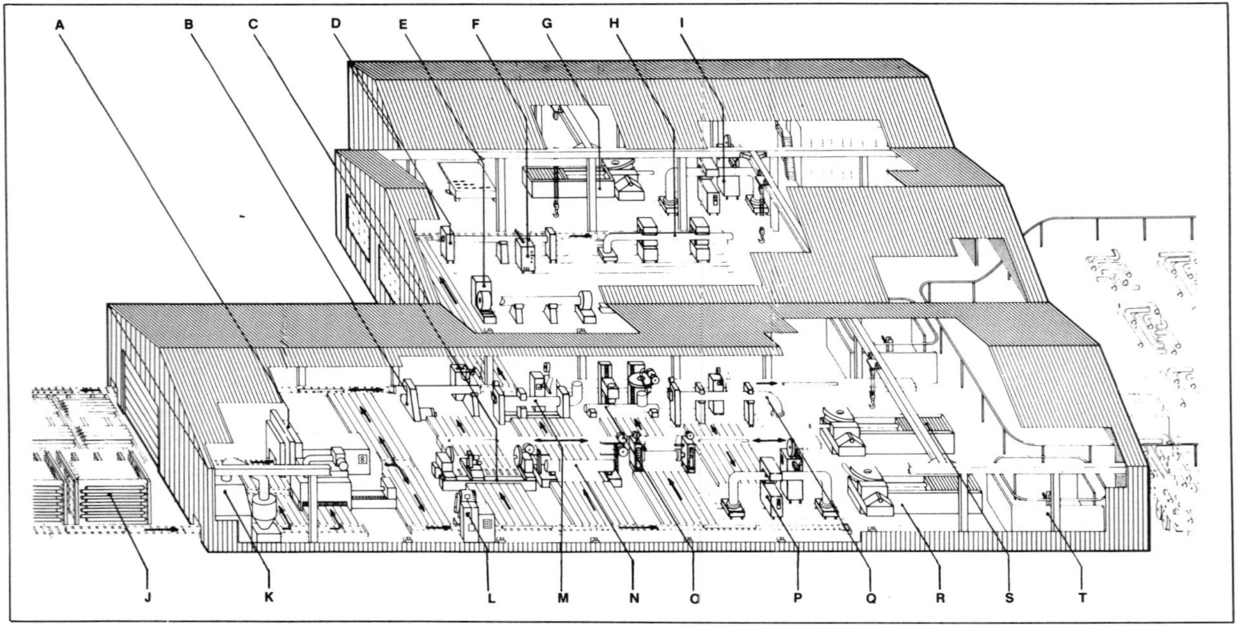

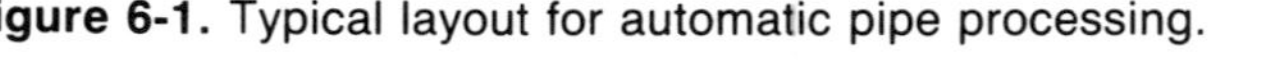

A-Pipe-coating machines provide internal and external painting systems, including drying oven.

B-Contour cutting machine for 2 in. to 12 in. pipe combines contour cutting and beveling of various alloys of pipe.

C-Automatic edge-preparation machine allows for automatic beveling of either or both ends of pipe prior to fitting and welding.

D-Automatic pipe flange tacker and welder with double rotator positioners for flange fitting and welding of 24 in. to 36 in. pipe.

E-Contour cutting machine (gas or plasma) for contour cutting and beveling of 16 in. to 24 in. pipe.

F-Manipulator holds welding gun in position during semi-automatic welding process.

G-Bending machine for 12 in. pipe is standard three-dimensional bending machine.

H-High boy with positioner and dolly provides positioning of pipe for fitting and welding.

I-High boy with positioner, support dollies, and manipulator provides positioning and welding of large-diameter pipe.

J-Automatic pipe storage rack , for 2 in. to 24 in. pipe, provides automatic loading, selecting, and off-loading to transfer system.

K-Internal and external blasting machines provide internal and external blasting for surface preparation of pipe.

L-Measuring, cutting, and marking machines measure and cut pipe, locate holes, mark pieces, and identify pipe.

M-High boy rotator with two variable dollies, a semi-automatic manipulator for fitting, tacking, and welding pipe elbows.

N-Automatic flange fitting and welding machine selects and orients flanges, tacking and welding inside and out.

O-T-drill and branch welder, used for extrusion of branch joint collars and automatic welding of 90° branches.

P-High boy with two positioners and dolly provides positioning of pipe for fitting and welding.

Q-Semi-automatic high elbow welder manipulator for fitting, tacking, and welding long-radius elbows.

R-Bending machine for 4 in. to 10 in. pipe automatically bends flanged pipe in three dimensions by numerically controlled instructions.

S-Bending machine for 1 in. to 4 in. pipe automatically bends flanged pipe in three dimensions by numerically controlled instructions.

T-X-ray booth utilized to assure high-quality welding.

Figure 6-1. Typical layout for automatic pipe processing.

Figure 6-2. Silo storage for pipe. (Courtesy of Oxytechnik.)

Figure 6-3. External cleaning (shot blasting) stations. (Courtesy of Oxytechnik.)

system to the end preparation station where it is beveled (Figure 6-4). Oxytechnik provides a unit that will quickly and accurately bevel to say 37½°, while maintaining a constant land even though the pipe itself may be slightly out of round. This is very critical in a mechanized operation. Otherwise one would have to check each piece and very likely rework those with imperfectly beveled ends as is often the case with some mechanical methods.

Having now a cut length correctly beveled at each end, once more the pipe is conveyed onto the next station, usually a flanging stage. For slip-on flanges beveling the pipe may not be a requirement, in which case it would either by-pass the end prep stage or be processed straight through without any work being performed. However, let's just suppose it has been beveled for weld-neck flanges.

Figure 6-4. End preparation (beveling) station. (Courtesy of Oxytechnik.)

Once again, the pipe automatically has a flange welded on it at each end, in three distinct steps: fit, tack and weld.

Here again the computer does all the necessary figuring to insure that the flange holes are positioned correctly, not only for straight run lengths, but also and especially for lengths which are to be bent either on a single plane or even more critical yet, for those that will have multiple bends and planes. See Figure 6-5.

At this point the system can either stop or it can be designed to store the completed lengths, move them to other manual work stations or bending areas or have even further processing stages added. Each system is designed and tailored to meet a particular customer's requirement and no two systems are ever identical, just as no two shop areas are ever identical. Whatever the case, the automated system has a lot going for it.

Figure 6-5. Flange holding and positioning equipment for automatic slip-on flange welding. (Courtesy of Oxytechnik.)

As an example, a 4″ Schedule 40 carbon steel pipe spool, 10 ft in developed length, with a flange at each end and three welded corners, would normally take something like 405 minutes to fabricate using four lengths of cut and beveled pipe, three butt-weld elbows and two flanges. The same size and configuration spool processed through the automated system for a bend fabrication would require only the two flanges and a single length of cut and beveled pipe and would be processed in something like 15 minutes.

Obviously, the automated system saves considerable time, money, and manpower.

There are many other time, cost, and labor-saving processes associated with bending in pipe spool fabrication, such as branching and Van Stone flanging, both of which can be incorporated into an automated system, although there are limitations.

Branching

There are several methods for branching, or pulling a neck out of a straight section of pipe to eliminate the need for a T-fitting and three welds. Perhaps the best known equipment is T-Drill, which originated in Scandinavia and was designed initially for small-diameter copper tubing. The method is relatively simple. A hole is drilled in the pipe and a special tool is passed through the hole and expanded inside. As the tool rotates it gradually forces the material upward into a neck. When the neck is fully formed, the tool is withdrawn and the neck is squared and beveled. Until recently this process applied to thin wall material only, although in terms of size the upper limit was 18″ OD. Then T-Drill developed a method of heating the area to form collars in pipe up to Schedule 40 in wall thickness. The company is presently working on a method for heating the pipe by induction rather than flame, since the latter method tends to create a few problems in protecting the equipment itself from the torch heat.

There are, as I mentioned, other methods, particularly for heavy wall material, whereby a mandrel-type ball is used to pull a neck out of the heated area.

Van Stone Flanging

Lap joints for use with slip-on flanges have long been used in pipe fabrication, but until A-587 carbon steel was developed they had to be fabricated and then welded onto the pipe ends if they were to be in carbon steel. The main advantage to using this method of joining two pipe sections is that the flanges are loose, so they are easily rotated to align the holes, and simply bolted together. Unfortunately, while the machinery was developed to form a lap joint flange at the end of a piece of pipe that worked well with stainless steel and other materials, none of the usual carbon steels such as A-53 and A-106 would accommodate the severe 90° bend to cold form the lap. They always cracked. Then A-587 carbon steel pipe was developed which is considerably more ductile and this is now what is used almost exclusively for Van Stone flanging. Interestingly, A-587 material has another distinct advantage. Because of a finer grain and lower carbon content than either A-53 or A-106 or the other carbon steel pipes such as A-120 and API-5L, it is easier to bend and is much more controllable in terms of wall thinning and ovality. In fact, 1½D bends in this material may have as little as 8% wall thinning, whereas the other materials seldom offer less than 16%.

The cold mechanical process is quick and simple to operate. Most machinery today can accommodate up to 8″ pipe, although A-587 is not presently available in anything larger than 4″ Schedule 40.

If one goes to a hot process, of course, lap joint flanges can be made in sizes up to 12″ or even larger, and there is a German induction machine available that will handle up to 18″ pipe. It is relatively quick, but somewhat expensive.

Apart from bending, per se, which eliminates the need for two welds at every corner in a piping system, only the lap joint flanging and the collaring equipment provide further means of reducing welding requirements. All other ancillary and support equipment, while providing labor-saving steps in pipe fabrication, contribute little or nothing to reduction or elimination of the welding aspects. Almost every shop will have a variety of welding machinery, positioners, manipulators, etc. in addition to whatever bending, material handling, processing and testing equipment they may have.

7

Piping Design and Fabrication

For years, piping has been designed and fabricated along highly standardized, conventional lines. The normal method for turning a corner in a piping run has been to use a standard butt-weld fitting, which calls for a 1½D or long radius elbow and two lengths of pipe for the same diameter and schedule wall thickness. The straight pipe must be cut to the required length and the ends beveled in preparation for welding. The straight sections are then fitted into position on either side of the fitting, tack-welded into place and finally welded out. More often than not, when that has all been done the weld must also be x-rayed.

Naturally, all this takes time. It involves a number of different workers and clearly defined procedures to be followed in accordance with prescribed standards and codes.

For example, a pipe spool in 18″ OD, Schedule 60 carbon steel with 6 weld-ell corners requiring 12 welds would cost approximately $11,627 to fabricate manually, with a material cost of $8,600 and labor $3,027 (See Fig. 7-1).

By using 3D bends in lieu of 1½D weld-el fittings, while still maintaining critical center-to-center dimensions, the cost is re-

PIPE BENDS REDUCE COSTS

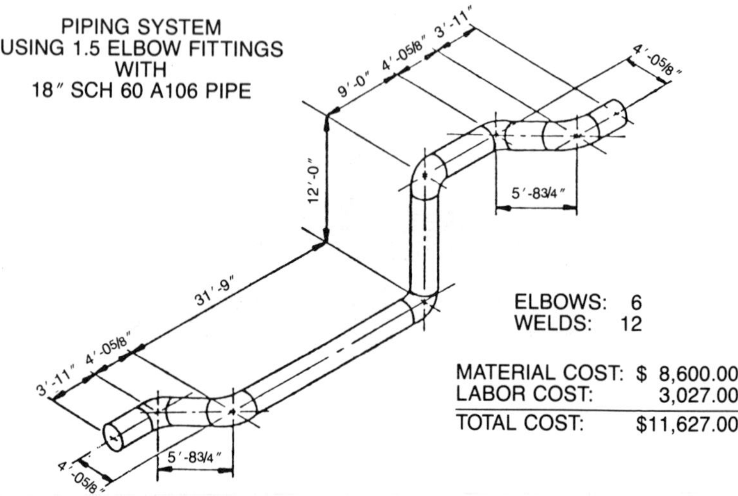

SYSTEM SAVINGS USING 3D PIPE BENDS:

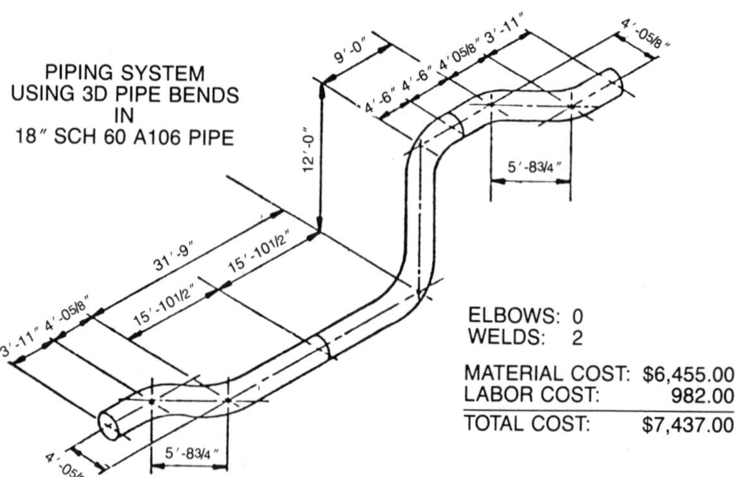

36% REDUCTION IN FABRICATION COST

35% REDUCTION IN FRICTION LOSS

Figure 7-1. Example of cost savings using 3D bends. (Courtesy of Associated Piping & Engineering Corp.)

duced to \$7,437. Material is now only \$6,455 and labor a mere \$982. The number of welds has been reduced to two (to join the three sections needed to make the spool) and the fittings have been eliminated.

This represents a cost reduction of 36%. Friction loss has also been reduced 35% by going to a 3D corner configuration instead of the original 1½D. Thus we have improved design to effect lower in-use cost and lowered the fabrication cost as well.

Effectiveness of Automated Production

In terms of cost, automated production can offer savings of 20–60% over conventional methods, depending upon the degree of automation provided and how far back the programming for automation originated. If the system is only partially automated then the savings will be lower, but if that same spool piece was computer-designed in the first place and the main frame was used to program the production equipment, the spool could be produced completely automatically without having to have any accompanying drawing.

One chemical company has already stated that for maximum cost effectiveness from all viewpoints: design, engineering, fabrication, installation, end use operation, and maintenance, it would like to move to total computerization. That, regretfully, is still not possible today, although soon it probably will be.

Design Standards

Almost all piping design is still constrained to conform to the standard 1½D weld-ell corner configuration, although it is generally accepted that this is not always or necessarily the ideal one. Theoretically, the optimum radius for any corner would vary slightly from one to the next, depending upon a number of factors, but making five corners in succession in a single spool piece of, say, 2.9D, 1.8D, 2.6D, 2.8D and 2.1D is somewhat pointless, even though those particular D figures may be as close to perfection as one can get. 1/10″ difference in radius is unlikely to affect friction loss or pressure drop factors, but a difference between 1.5D and 2.8D could. The problem is two-fold. First, no one can say for sure that a particular D is *the* ideal. Various sources have claimed that 3D or 4D or 2.5D is probably the best "standard", but to some extent the D determined may have been predicated upon one formula while another so called "ideal" has been based on a totally different formula.

In principle, the feeling is that somewhere in the region of 3D is close to general "optimum", but here again there is some measure of confusion because in terms of piping, neither 1½D nor, for that matter, any other specified D is a true D because the IPS sizes to 12″ are all nominal and 1½D is actually anywhere between 1.14D and 1.41D in pipe ranging from 1–12″ in size. Thus a formula based on true OD which shows that 2.66D would be the ideal is pretty close to another formula based on nominal size which says that 3D is ideal when applied to 4″ pipe. Both are using a 12″ radius, thus:

$$\frac{12}{4.5} = 2.66\text{D for } 4\tfrac{1}{2}'' \text{ true OD}$$

$$\frac{12}{4} = 3\text{D for } 4'' \text{ nominal pipe diameter}$$

A further complication, of course, is that it just isn't practical to keep switching radii in production. Apart from the enormous cost of tooling involved, there is also the inevitable downtime required to change the tooling, or at least the die and associated tooling settings. This particular problem can be overcome with induction bending, which can change a radius by as little as ¹⁄₁₀″ without any downtime or tooling cost involved. This type of machinery for small-diameter pipe is only just coming onto the market and is not yet in widespread use, although it does have far better wall thinning control capability. It will also undoubtedly get more attention because of its ability to achieve tolerances equivalent to those established for a fitting at 1½D and substantially better as the radius is increased.

There is also the fact that there simply is not enough bending data available. We have neglected adequate research and development in this area. Existing standards, specifications and codes are mostly based on hot slab bending technology or outdated cold bending machine capabilities which have been vastly improved over the past several years.

Without any published national standards for cold and induction bending, most piping designers and engineers are reluctant or unable to take any other approach than the one they use now—conventional weld fabrication.

Nevertheless, things are changing and the problem of standards for cold and induction bending is now being addressed in greater detail in a much more aggressive manner by the bending industry through a new industry organization formed early in 1983—the International Pipe Association.

Other Cost Factors

Apart from bending, which goes a long way toward effective cost reduction in pipe spool fabrication and which, with the use of better radii for improved flow characteristics, will provide lower operational costs through lesser energy (pumping) requirements, there are other factors that contribute to lower costs.

One of these, modular construction, is gaining growing favor and attention even though it is not a new method. Certainly by designing a plant and piping system in a series of sections or modules so that each one is prefabricated with all its valving, instrumentation, etc., and tested at the shop before it is shipped to the field, conventional installation costs can be reduced substantially.

Most shipyards today build their ships using the economical modular method. Some have already applied this technology to chemical plant construction. We will see a great deal more modular design and construction in the future.

Maintenance programs will be much more predictable through greater use of computers. At the time the system is designed, a great deal of data is accumulated. By entering much of this in a computer system, it will be possible to develop a long-term maintenance program that draws upon accumulated data to schedule inspections, shutdowns, repairs, and replacements when coupled with actual operating input. Between design data and operating data, the computer will be able to establish a very cost effective maintenance program.

Associated with maintenance, of course, are all the problems inherent with welding, corrosion, and erosion. It is not too far-fetched to think that one day we may be able to virtually eliminate welding from a piping system, particularly where ends are flanged rather than butt-welded. Even in the latter case, equipment could be developed for field application.

As an example of some of the problems in these areas, one company is replacing elbows at the rate of one per month because of stress corrosion in the weld area and internal erosion in the elbow itself. Technology exists, although it is not yet fully developed, that will (A) make a bend for the corner, thus eliminating the fitting and the welds on either side and yet maintain the same tolerances as for the fitting and (B) surface-harden the interior of the bend to provide for up to five times longer life. Alternatively, it is possible to eliminate the welds by going to electrohydraulic fusion of fitting and pipe ends with the fitting already surface hardened. One approach to the problem would be to improve the corner radius first, determine the true optimum for minimum friction, bend it to that radius and then surface harden the interior. As a result, instead of replacing elbows a dozen times a year, it might only be necessary to do so twice a year—at very considerable cost savings. Again, for replacement, if the tangent lengths have flanged ends, the simple solution would be to replace the whole piece, which becomes a matter of unbolting the flanges, taking out the worn corner section and rebolting the replacement section into position. The downtime would be far less than is presently the case and consequently be at much lower cost.

Electrohydraulic fusion is seen as a very positive step toward cost reduction in pipe bending and fabrication, particularly for exotic materials. Since any bending process necessitates having tangent lengths, in most cases two to three times the diameter of the pipe, and these may at times have to be trimmed back, fusion would permit use of only the required developed length of material for the bend plus whatever tangents are needed. Any additional lengths fused on only to make the bend would be trimmed off again once the bend was completed. The extra lengths could be of almost any material since dissimilar metals can be readily joined by the electrohydraulic fusion method. Thus, there would be no waste of the exotic material itself.

Similarly, cut scrap could be recycled into new lengths, whether plain carbon steel, stainless, high alloy, exotic, or nonferrous material. For those bending machines with rear feed capability, random lengths could be fused into one continuous length with bend sections cut off on the machine after the final bend has been completed.

Flying cut-off equipment on bending machinery is not new. For certain applications it is used regularly, but there is always the need to first load a single length. This can be eliminated with continuous feed.

Figure 7-1 shows a CNC bending machine with flying cut-off attachment, used in conjunction with empty bending, i.e. no mandrel. Thus, if it were possible, the machine could be pro-

Figure 7-2. CNC bending machine with flying cut-off attachment. (Courtesy of Schwarze-Wirtz A.G.)

grammed for use with continuously fed material, if 20′ straight lengths were electrohydraulically fused prior to entering the machine from the rear.

There is no doubt that we are entering new areas in piping design and fabrication and that computer-aided design in conjunction with computer-aided production is going to be the direction to follow. With such additional capabilities as automated material movement and processing, variable radius bending, internal hard surfacing, electrohydraulic fusion and stereoptic dimensional verification, we should be seeing some tremendous changes taking place in these fields.

Unfortunately, technology is out-pacing our ability to establish appropriate standards and specifications for some of these advances and we are sadly lacking in our approach to R & D.

The ills that have beset some of our industries in the past, such as automotive and steel production, can be blamed in part on our failure to grasp and take advantage of technological changes. When we hear that in Japan prefabricated housing comes off the production line every 40 minutes and can be field-erected in four hours with a 20 year warranty, we must look to achieving similar results in our own industries, particularly the piping fabrication field.

If a pipe spool can presently be fabricated in 15 minutes by automated production and bending as opposed to 405 minutes by conventional methods, think how much quicker yet fabrication could be accomplished by fusing components together rather than welding. Automated production really applies only to storage, cleaning, cutting, end preparation, flanging and bending corners. The other aspects such as welding tees, nozzles, and reducers are still mostly done manually. This is what takes the time in piping fabrication and this is what could be reduced substantially in terms of time and cost with a technology such as electrohydraulic fusion.

Automating our fabrication facilities will not happen overnight, but as companies see the advantages of incorporating these newer technologies and can finance them, our productivity will improve with concurrent lower costs and perhaps eventually get us back into a competitive world-wide position again.

8

Standards and Quality Assurance

Standards

Currently, such standards and codes as there are for pipe bending are limited to:

A. Hot slab bending
B. Certain aspects of cold bending

The Pipe Fabrication Institute (PFI)—standard ES-24 (April 1975) covering pipe bending tolerances (PBS)—minimum bending radii (MBR)—minimum tangents (MT)—clearly states "All PFI standards are advisory. There is no agreement to adhere to any PFI standard and their use by anyone is entirely voluntary."

Nevertheless, much of the detail in ES-24 has been used as an industry standard for many years and is incorporated in certain codes. Among the data contained in the standards, which are based entirely on the hot slab bending methods, ovality in a bend is required to be not more than 8% at radii of 5D or greater, when the ratio of *nominal* diameter to *nominal* wall is 35 or less. Thus, a 24″ OD pipe with a wall thickness of .969″ (Schedule 60) would have a 24.76 ratio factor—well within the 35 limit. Bent 90° on a 6D radius (144″), the ovality must not exceed 8%. That sounds quite reasonable and for hot slab bending, it is. But compare that percentage to the *predicted* amount of wall thinning for the same piece of pipe bent by the induction method, where the ovality will be 1.38%—possibly less if the machine is run by a skilled operator. At least one manufacturer uses that figure as a standard for all its induction machines. There are almost a dozen of these machines operating in the United States today.

Because wall thinning must also be taken into consideration in bending, the PFI Standard has a formula that specifies "minimum" wall thickness of the pipe at certain radii *prior* to bending so that after bending is completed, the wall thickness will not be less than that for the straight pipe.

At 6D this is 1.06 tn where tn is the minimum calculated wall thickness as required by the applicable code. Now, if the 24″ OD pipe we are using as an example is, say, A-106, the specification for this states that the wall thickness at any point shall not be more than 12.5% under nominal. API 5L pipe provides for plus 15% and minus 12½% for seamless pipe and plus 15% minus 10% for welded pipe.

Using the API formula for a wall thickness of .969″ as nominal and multiplying it by 1.06, we get a figure of 1.02714″ as the recommended thickness for bending in order to keep the wall thinning within the specified minimum tolerance of not more than 12½% less than nominal in the straight material, for both A-106 and API 5L. Thus provision is made for something like 18% wall thinning if the normal specification for the straight pipe is applied as the minimum thickness permissible *after* bending.

However, going back to the computer which is used to determine the predictable wall thinning by induction, we find that the percentage of wall thinning for a 6D bend in 24″ OD × .969″ wall carbon steel pipe will be 6.25%. Consequently, there is no need for applying the PFI formula if the thinning *after* bending is only 6½% when the manufacturing tolerance itself provides for 12½%. In fact, by ultrasonically checking the pipe wall prior to bending to determine where it is heaviest one could very easily end up with a wall thickness greater than the nominal for API 5L pipe which allows for a plus figure of 15%.

One can see, therefore, a need to establish some standards other than those predicated on hot slab bending for induction bending only. The situation is like asking us to drive our cars today under the rules that prevailed when a man with a flag had to walk in front of a motor car. We are being held back by our failure to keep our standards in line with our technology. Induction bending has been applied in the United States now for nearly a decade and in Europe for over 15 years. And here we are still without any national standards for this process which is in use throughout the world today.

Cold Process Standards

Here we run into all sorts of formulas. The American National Standards Institute—ANSI—has established codes for pressure piping which fall under several different sections of B.31 with B.31.3 written specifically for chemical plants and refineries.

To start with, section 329.2 Procedures states that "Pipe may be bent by any hot or cold method permissible by radii and material characteristics of the pipe being bent." No distinction has been made at all between the two types of bending (three, really, if we see hot slab and induction as separate technologies—which they are).

When we look at wall thinning, section 304.2.1 states that "the minimum required thickness, tn, of a bend, after bending, shall be determined as for straight pipe in accordance with 304.1."

According to 304.1, "the required thickness for straight sections of pipe shall be determined in accordance with equation 2: tm = t + c and that "the minimum thickness for the pipe selected, considering manufacturer's *minus tolerance,* shall not be less than tm."

tm = minimum required thickness, including mechanical, corrosion and errosion allowances.

t = pressure design thickness, as calculated in 304.1.2 for internal pressure or in accordance with the procedure listed in 304.1.3 for external pressure.

c = the sum of the mechanical allowances (thread or groove depth) + corrosion and erosion allowances.

Thus the minimum acceptable wall thinning for bending is predicated upon a number of different factors which are determined by the design engineer who may well decide that under certain conditions which he has calculated, wall thinning of 18% or 22% may be acceptable.

A number of chemical companies, in fact, have established their own specifications for bending. While they are generally all similar to each other, there are some differences. One company has even gone so far as to have two sets of specifications—one for in house bending and the other for subcontract outside benders.

Another company states in its specifications "centerline radius of pipe bends shall be equal to at least five times the nominal pipe diameter" and that "pipe bends shall be used for small diameter pipe in place of welding elbows, unless impractical." But then it goes on to say "pipe bends shall not be used for general piping where weld fittings can be used" and adds "when bends are used increased wall thickness will be adequate for intended use."

Talk about confusion and ambiguity! If that particular company was, for instance, designing a piping system using A-587 material, cold bending would, for 1½D bends, get wall thinning in the range of 8–9% only, as opposed to butt-weld fittings which have a permissible wall thinning tolerance of not more than 12½%. Bends, therefore, would be more appropriate.

Yet another company spells out their bending specification which limits the amount of flattening, thinning, wrinkling, and necking permitted for cold bending to no more than 6″ IPS. No mention is even made of either hot slab or induction bending. Bend radii of 1½D, 3D and 5D are provided for in accordance with certain materials, and for 1½D and 3D bends wall thinning of 18% is allowed, reduced to 10% for 5D. Ovality at 1½D allows a maximum of 5% for Schedule 40 and heavier in pipe under 6″ and 8% for 3D and 5D with 7.3% for 6″ Schedule 40 material only in the same two D's of bend.

Generally speaking, among all the companies that have prepared their own specifications for bending, up to 18% wall thinning is acceptable for 1½D and 3D bends and up to 10% for 5D.

Some even go so far as to specify the amount of wrinkling that is acceptable on the inside of a bend. This is based upon PFI Standard ES-24 which delineates wrinkling or wave tolerances as follows:

1. All wave shapes shall blend into the pipe surface in a gradual manner.
2. The maximum vertical height of any wave measured from the average height of two adjoining crests to the valley shall not exceed 3% of the nominal pipe size.
3. The minimum ratio of the distance between crests as compared to the height between crests and the included valley shall be 12:1.

One company specifies depth in accordance with pipe size as:

> The depth of wrinkling from crest to trough shall not exceed the following:

Nominal pipe size up to	2″	3″	4″	6″
Depth	1/32″	3/64″	1/16″	3/32″

Provision is also made for necking. The stipulated tolerance for this is that the reduction in *circumference* in the neck area shall not exceed 4% of the *nominal circumference* (*not* diameter).

One of the problems that affects bending and consequently any standards is that such codes, specifications and standards as there are tend to categorize pipe simply as pipe and bending as bending. It would be much more appropriate if standards were based on (a) each of the different types of pipe and (b) each of the different types of bending.

Tolerance Variables

There are too many variables in pipe manufacturing dimensional tolerances to be able to have one general standard. For example, cold bending A-106 pipe will generally not get much better than 16–17% for wall thinning, whereas in A-587 12½% is readily achievable and has been reportedly as low as 8%. Tooling in cold bending plays a very critical role and too tight a fit of the pipe and the tooling or too loose a fit and/or the same thing for the mandrel fit, can seriously affect the quality of the bend and bend tolerances. The OD tolerance for A-53 is the same as for A-120 material—±1% for 2″ and larger, whereas for A-106 the OD tolerance goes from 1/64″over for 1½″ and smaller to 3/32″ for 10–18″ and 1/8″ for 20–24″ pipe with an across the board 1/32″ under for all sizes.

Despite these inconsistencies, there is no reason parameters cannot be set for each category of pipe and for each type of bending in all sizes within the normal range of capabilities. These parameters could then be developed into national standards which would cover the whole range of piping in complete detail such as wall thinning, ovality, fiber elongation, wrinkling, and necking at different radii. With clearly defined bending standards, every piping designer and engineer would then have the necessary information at his fingertips to determine whether he wanted his system fabricated conventionally or by bending and would be able to make his choice accordingly.

Quality Assurance

For decades, quality assurance and dimensional verification of bends was very often a matter of "if it looks good, it is probably all right." For small diameter materials, pipe jigs or fixtures were often used and if the workpiece fitted, it was good enough to pass. For larger material, especially pipe, sometimes a fixture would be used, but more often than not, a chalk-line floor layout.

Neither method was reasonably accurate, and could never really satisfy close critical tolerance requirements as far as dimensions were concerned. Similarly, loss-of-ovality measurement was usually done, and still is in many places, by means of a large pair of calipers and a ruler.

Wall thinning, of course, was determined by sectioning the pipe, which was fine as far as getting a measurement was concerned, but only for that particular bend, which having been sectioned, was then a total loss.

The advent of ultrasonic equipment resolved the problem of wall thickness measurement and today it is used extensively, both before bending to determine where the thickest part of the wall is in order to get that area on the outside of the bend, as well as after the bending to see exactly how much thinning has taken place and where. Computer technology also solved the introduction of a much more precise method of measuring workpiece configuration accuracy, particularly for small material. NC machinery for tube bending proceeded satisfactorily with effective methods of product quality assurance. But for a while there was considerable dissatisfaction with NC and CNC bending because of the inability of support equipment to verify results to an acceptable degree.

Among the machinery manufacturers addressing themselves to the problem in the United States, the first was Eaton-Leonard Corporation. They developed the "Vector-System," which describes a tube shape, not by its bend configuration, but by a series of intersecting lines or vectors. In the early sixties, the United States automotive industry initiated X-Y-Z coordinates for the purpose of dimensioning tube drawings. Eaton-Leonard recognized that these coordinates defined vectors, so they developed a system to bend, measure, and inspect tubes accordingly. Figure 8-1.

Basically the system, used in conjunction with a compatible bender, not only measures and inspects the end product, but also corrects for the many process variables such as tool wear and adjustment, material properties, surface finish, lubrication, and pressure settings. While these variables remain fairly constant during a normal production cycle, they can change, and therefore must be monitored by periodic checks of the workpieces. Certainly, between shifts, if the production run is extensive,

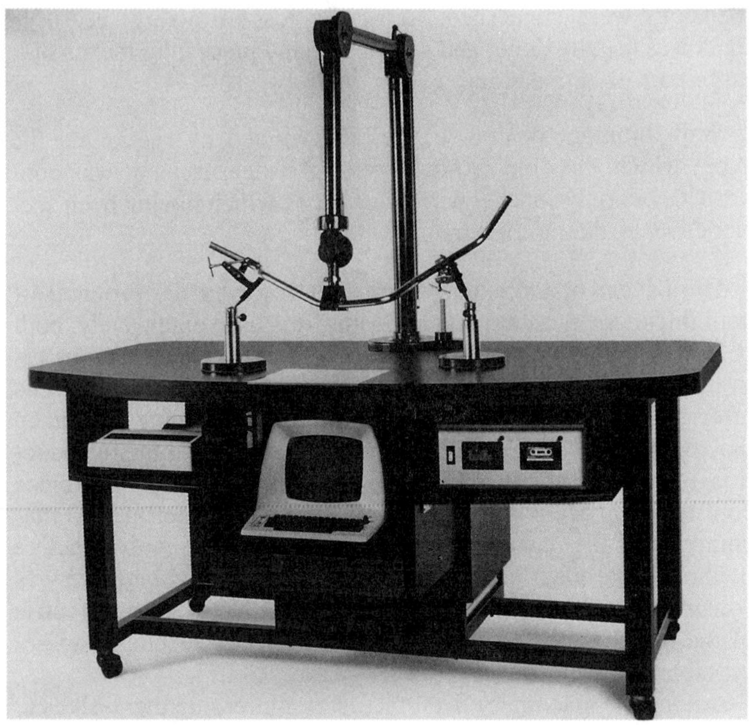

Figure 8-1. Vector System digitized tube data center. (Courtesy of Eaton-Leonard Corp.)

changes can occur, thus the first and last pieces must always be checked for every run.

The Vector System and others like it, such as the Tracemaster, developed by Schwarze-Wirtz in Germany, are ideal for smaller diameter materials (Figure 8-2). For large diameter piping, where dimensional verification of a complex spool piece in, say, 24″ OD by 2″ wall pipe, may be necessary, other methods are used such as laser beam measurement. Even so, the workpiece must usually be positioned before any measurement can take place and this in itself is both time-consuming and labor intensive.

Very recently an aerospace technology, such as is used to measure the F-16 fighter plane, has been adapted for dimensional

verification of pipe spools without the need for any positioning of the pipe.

This is the AIMS (Analytical Industrial Metrology System) developed by the Keuffel and Esser Co., which measures the XYZ coordinates in three-dimensional space utilizing two digital theodolites as a data source connected to a micro-processor and printer (Figure 8-3).

By using small measurement fixtures that are readily set onto or attached to the pipe to provide aiming targets for the theodolites—the rounded surfaces of pipe and bends do not offer sufficient accuracy for targeting—all the critical dimensions such as tangent points, center-to-face and face-to-face distances, radius, degree of bend, plane of bend, distance between bends and even

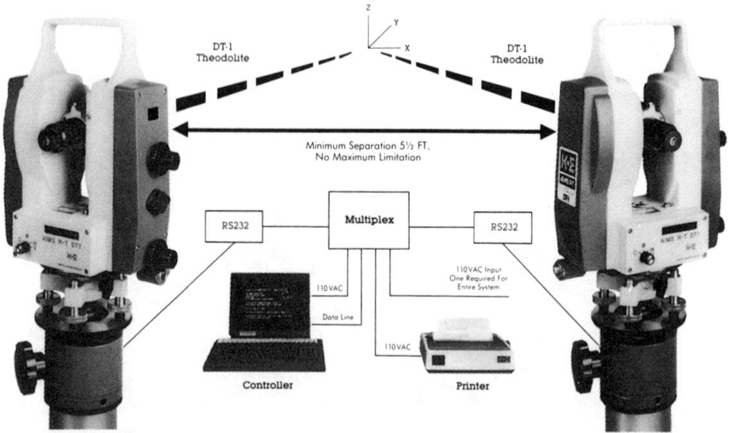

Figure 8-3. Components of the AIMS System. (Courtesy of Keuffel and Esser Co.)

ovality are very quickly determined, accurate to 1/1000″ or degree with, if required, a hard-copy printout.

One big advantage to this system is that the workpiece does not have to be positioned before measurements are taken. It can be placed in whatever way it sets after being unhooked from the overhead crane that moved it to the station.

Another advantage is that the equipment is portable and can be moved to a field jobsite to determine critical dimensions of a pipe spool due to be replaced. By measuring the spool accurately while it is in place the resultant data can then be transmitted to the job, programmed into the fabrication and/or bending equipment to produce an exact duplicate of the spool with verification, of course, as to dimensional accuracy by the AIMS System. Thus downtime is limited to changing over the two identical pieces.

Such support equipment is now providing far greater accuracy in quality control, in far less time with minimal labor input and, consequently, lower cost.

APPENDICES

Glossary of Terms

AGE HARDENING — See heat treatment.

AIR FRAME TUBING
This tubing is produced for aircraft structural parts. This tubing is made to special surface quality, mechanical properties and other characteristics required by Military Specifications (MIL-T-. . .) and SAE Aeronautical Materials Specifications (AMS . . .).

AIR HARDENING — See heat treatment.

AIRCRAFT QUALITY
Is a steel which has a special cleanliness rating determined by magnetic particle testing. The terms "Aircraft Quality" and "Magnaflux Quality" are considered synonymous.

ALLOY STEEL
All steels contain carbon and small amounts of silicon, sulfur, manganese and phosphorus. Steels which contain intentional additions of elements other than these, or in which silicon and manganese are present in large amounts for the express purpose of improving or altering any of the physical or mechanical properties of the steel, are termed alloy steels.

ANNEALING — See heat treatment.

AUSTENITIC STAINLESS STEEL
Low carbon, iron-chromium-nickel stainless alloys containing more than 16% chromium, with sufficient nickel to provide an austenitic structure at normal temperatures. These alloys cannot be hardened by heat treatment, but can be hardened by cold working. They are normally non-magnetic, but can be slightly magnetic depending upon composition and amount of cold working.

AVERAGE WALL — See dimensions.

BEARING QUALITY STEELS
Steels suitable for use in balls, rollers and races of high quality anti-friction bearings.

BEVEL
An angular cut on the I.D. or O.D. of a tube end.

BILLET
As used in the manufacture of seamless tubes, a round bar with dimensions and other characteristics suitable for piercing into tubing.

BLISTER
A raised spot on the surface of the metal caused by expansion of gas in a subsurface zone during thermal treatment.

BLOOM
A semi-finished piece of steel, resulting from the rolling or forging of an ingot. A bloom is square or rectangular not more than twice as wide as thick, and usually not less than 36 sq. in. in cross-sectional area.

BORESCOPE
An optical device used for inspecting under low magnification the inside surface of tubes.

BRIGHT ANNEAL — See heat treatment.

BRINELL HARDNESS — See hardness.

BROKEN SURFACE
A surface having innumerable minute cracks running normal to the direction of working.

BUS CONDUCTOR
A rigid electric conductor of any cross section, i.e., Aluminum Bus Pipe.

CAMBER
The amount of curvature or deviation from exact straightness over any specified length of tubing.

CAPPED STEEL
Semi-killed steel which has characteristics similar to those of rimmed steels but to a degree intermediate between rimmed and killed steel. The capping operation limits the time of gas evolution and prevents the formation of an excessive number of gas voids within the ingot.

CARBIDE
A compound consisting of carbon and other elements.

CARBIDE PRECIPITATION
The phenomenon of carbides coming out of a solid solution, occuring in stainless steel when heated into the range of 800-1600 degrees Fahrenheit.

CARBON STEEL
A steel consisting of essentially iron, carbon, manganese, and silicon. Carbon steel has no minimum content required for aluminum, chromium, cobalt, columbium, molybdenum, nickel, titanium, tungsten, vanadium, zicronium or any other element added to obtain alloying effect. Small quantities of certain residual elements are considered incidental.

CARBURIZING
Adding carbon to the surface of iron-base alloys by heating the metal below its melting point in contact with carbonaceous solids, liquids or gases. Desired hardness and toughness properties are developed in the high carbon "case" by quenching and tempering.

CASE HARDENING
A heat treatment in which the surface (case) of an iron-base alloy is made harder than the interior (core). Any of the following methods may be employed: flame hardening, induction hardening, carburizing, cyaniding or nitriding.

CHARPY IMPACT TEST — See impact testing.

CHECK ANALYSIS
An analysis of the metal after it has been rolled or forged into semi-finished or finished forms. It is not a check on the ladle analysis, but is a check against the chemistry ordered, i.e., Product Analysis.

CHLORIDE STRESS CRACKING — See Stress Corrosion Cracking.

CLEANUP
The amount of metal removal required to obtain desired dimensions and complete removal of inherent surface imperfections.

COEFFICIENT OF THERMAL EXPANSION
A physical property value representing the change in length per unit length, the change in area per unit area or the change in volume per unit volume per one degree increase in temperature.

COLD DRAWING
A process in which tubing is drawn at room temperature through a die and over a mandrel to achieve its final size and to provide better surface finish, closer tolerances, lighter walls, smaller diameters, longer lengths, or a different combination of mechanical properties from those possible through hot finishing or direct welding.

COLD REDUCTION
The reduction of sectional dimensions of a tube by any of a number of types of cold-working operations.

COLD SINKING
Similar to cold drawing, except that the tube is drawn through a die, but without an internal mandrel. Usually used only for making heavy wall or small tubing, where drawing over a mandrel is impractical. Only outside diameter is closely controlled.

COLD WORKING
Permanent plastic deformation of a metal below its recrystallization temperatures.

CONDITIONING
The removal of surface defects (seams, laps, pits, etc.) from steel. Conditioning is usually done when the steel is in semi-finished condition (bloom, billet, slab). It may be accomplished, after an inspection, by chipping, scarfing, grinding, or machining.

COPPER-COPPER SULFATE TEST
An intergranular corrosion test for stainless steels. The specimen is placed in boiling copper-copper-sulfate-sulfuric acid for 24 hours after which it is bent to expose any surface intergranular attack. This test is often preferred over the Huey test because it requires much less time.

CORROSION
Chemical or electrochemical deterioration of a metal or alloy.

Galvanic Corrosion — Corrosion associated with the presence of two dissimilar metals in a solution (electrolyte). In principle, it is similar to bath-type plating in the sense that the anode surface has lost metal (corroded).

Intergranular Corrosion — Corrosion which occurs preferentially along the grain boundaries of the alloy.

Pitting Corrosion — Non-uniform corrosion usually forming small cavities in the metal surface.

CORROSION RESISTANCE
The ability to resist attack by corrosion.

CREEP STRENGTH
The constant nominal stress that will cause a specified quantity of creep in a given time at a constant temperature. It is a measure of a tube's ability to withstand prolonged stress or load without significant continuous deformation. In steels it is an important factor only at elevated temperatures.

CROWN
Crown, in plates, sheet or strips, is characterized by a greater thickness in the middle than at the edges. It may be caused by a deflecting (bending) of the rolls or by worn rolls.

CUT LENGTH
Refers to tubing ordered to a specified length and permitting a tolerance of a standardized fraction of an inch over but nothing under the specified length.

CYANIDING
A process in which an iron-base alloy is heated in contact with a cyanide salt so that the surface absorbs carbon and nitrogen. Cyaniding is followed by quenching and tempering to produce a case with a desired combination of hardness and toughness.

DECARBURIZATION

The loss of carbon from the surface of an iron-base alloy as the result of heating in an environment which removes the carbon. In medium or high carbon steels, decarburization leads to a pronounced lowering of the fatigue limit.

DENSITY

The mass per unit volume of a substance, usually expressed in the tubing industry in pounds per cubic inch.

DIE LINE

A longitudinal depression or protrusion formed on the surface of drawn or extruded material due to imperfections on the die surface.

DIMENSION

O.D. — Outside Diameter. Specified in inches and fractions of an inch, or inches and decimals of an inch.

I.D. — Inside Diameter. Specified in the same units as the O.D.

Mean Diameter — The average of two measurements of the diameter taken at right angles to each other.

Wall — Wall Thickness or Gage. Specified in either fractions or decimals of an inch or by a "wire gage" number. In the United States, the most common gage used for tubing is the "Birmingham" iron gage, designated "B.W.G".

Nominal — The theoretical or stated value of the O.D., I.D., or wall dimension as specified by the customer.

Maximum and Minimum — The dimensions resulting after applying the proper tolerances to the nominal dimensions.

Minimum Wall — Generally, the lightest wall permitted within specified tolerances. A "minimum wall tube" is one whose wall thickness is not permitted to fall below the specified nominal measurement.

Average Wall — A tube whose wall thickness is permitted to range over and under the specified nominal wall measurement within certain defined tolerances.

Mean Wall — The average of two measurements of the wall thickness of a tubular product taken opposite each other.

DUCTILITY
The ability of a tube to deform plastically. Frequently, elongation during tensile testing is used as a measurement of this property.

DYE PENETRANT INSPECTION
Non-destructive test employing dye or flouresent chemical and sometimes black light to detect surface defects.

ECCENTRICITY
The displacement of the I.D. of the tube with respect to its O.D., i.e., Deviation from a common center. The permissible degree of eccentricity can be expressed by a plus and minus wall-thickness tolerance.

EDDY CURRENT
Non-destructive testing method using eddy current flow for the purpose of recognizing a discontinuity in the piece being tested.

ELASTIC LIMIT
A measure of the maximum stress that may be applied to a tube without leaving a permanent deformation or strain after the stress is released.

ELECTRICAL CONDUCTIVITY
The capacity of a material to conduct electric current. For aluminum, this capacity is expressed as a percentage of the International Annealed Copper Standard which has a resistivity of 1/58 ohm-mm^2/metre at 20 C and an arbitrarily designed conductivity of unity.

ELECTRIC FURNACE PROCESS
One of the common methods used for melting and refining stainless and some alloy steels. It involves the use of electric power as the sole source of heat, thereby preventing contamination of the steel by impurities in the fuel as in other melting processes.

ELECTRIC RESISTANCE WELDED STEEL TUBE
Tubing made from strip, sheet or bands by electric resistance heating and pressure, the strip being part of the electrical circuit. The electric current, which may be introduced into the strip through electrodes or by induction, generates the welding heat through the electrical resistance of the strip.

As Welded Hot Rolled — ERW tubing exhibiting the pickled or shot blasted surface of hot rolled strip.

As Welded Cold Rolled — ERW tubing exhibiting the surface of cold rolled strip.

As Drawn tubing is unheat-treated, cold drawn tubing and has a scale free cold drawn surface.

Bright Annealed — Welded tubing normalized in a controlled atmosphere furnace and which exhibits a bright surface.

Pickled tubing has had the scale from hot fabrication or heat treatment removed by one of several types of acid solutions.

Gun Metal Finish — Welded tubing normalized, annealed, or stress relieved in a controlled atmosphere furnace which exhibits a gun metal finish.

Flash-In tubing is welded tubing which still retains the I.D. bead or flash formed during the welding operation. It can be furnished in either the as-welded, sunk or heat-treated condition.

Flash-Removed — Welded tubing from which the I.D. flash formed during the welding operation has been removed by some mechanical method. It can be furnished in either the as-welded, sunk or heat-treated condition.

Special Smooth I.D. — A cold drawn tube in which special attention is paid to the internal surface. Depth of pits and scores in I.D. are guaranteed to be below published maximum depths. Microinch finish is guaranteed in ERW tubes.

ELECTRICAL RESISTIVITY

The electrical resistance of a body of unit length and unit cross-sectional area or weight. The value of 1/58 ohm-mm meter at 20 C is the resistivity equivalent to the International Annealed Copper Standard (IACS) for 100 percent conductivity.

ELONGATION

The amount of permanent stretch, usually referring to a measurement of a specimen after fracture in a tensile test. It is expressed as a percentage of the original gage length.

ENDURANCE LIMIT

The maximum stress below which a material can presumably endure an infinite number of stress cycles.

ETCH TEST
Exposure of a specimen to acid attack for the purpose of disclosing the presence of foreign matter, defects, segregation pattern, or flow lines.

EXTRUSION SEAM
A seam in aluminum tube, pipe or hollow shape resulting from the pressure bonding at two or more edges in the course of extruding through a bridge/porthole die.

FATIGUE LIMIT — (Synonymous with Endurance Limit)

FERRITIC STAINLESS STEELS
The designation used for certain straight chromium steels which exhibit microstructures consisting mainly of ferrite at ordinary temperatures. Ferritic stainless steels are divided into two classifications; hardenable, and non-hardenable. When rapidly cooled from elevated temperatures the non-hardenable grades (ferritic) have a ferritic microstructure. The hardenable grades (martensitic) will exhibit a martensitic microstructure when rapidly cooled.

FINISH
Refers to the type of surface condition desired or existing in the finished tubular product.

FINISH ANNEAL — See heat treatment.

FINISH MACHINE SIZE
Normally specified in terms of the maximum machined O.D. and the minimum machined I.D. as applied to tubular parts. Finish machine size represents the size of the part as it comes

FINISH MACHINE SIZE
from the final machining operation. From this size the tube mill can calculate a tube size which will be guaranteed to cleanup upon machining.

FLAME HARDENING
A process of heating the surface layer of an iron-base alloy above the transformation temperature range by means of the flame of a high temperature torch, followed by quenching.

FLANGED END
In a flanged end the tube has been belled or expanded and a flange turned over until the wall of the tube end is at right angles to the wall of the tube.

FLASH-IN TUBING
See Electric Resistance Welded Tubing.

FLASH REMOVED — See Electric Resistance Welded Tubing.

FLUX LEAKAGE TEST
Non-destructive test which uses magnetic lines of force to recognize any discontinuity in the test piece.

FORGING
Used as a general term to describe the rolling, pressing or hammering of steel which displaces the metal under compression by a locally applied force, usually at hot working temperatures.

FRACTURE STRENGTH
As usually related to the tensile test, fracture strength or true breaking strength is defined as the load on the specimen at the time of fracture.

FULL ANNEAL — See heat treatment.

FULL FINISHED
Refers to stainless tube in which the weld has been processed to produce uniform strength and dimensions, and subsequently annealed to obtain proper corrosion resistance.

GAGES, GAUGES
A measurement of thickness. There are various standard gages such as United States Standard Gage (USS), Galvanized Sheet Gage (GSG), Birmingham Wire Gage (BWG).

GRAIN SIZE
A measure of the size of individual metallic crystals usually expressed as an average. Grain size is reported as a number in accordance with procedures described in ASTM grain size specifications.

Apparent Ferrite Grain Size is the average of the size of the ferrite grains as microscopically viewed in the normalized or annealed condition.

Austenitic Grain Size, which is usually measured by the McQuaid-Ehn method, represents the austenitic grain size of a material at a prescribed temperature above the upper critical, frequently 1700 F. For austenitic stainless steels the grain size does not change upon cooling and is that observed microscopically at room temperature.

HARDENABILITY
The property in steel that determines the depth and distribution of hardness induced by cooling from a suitable elevated temperature. The hardness can vary with the cooling rate.

HARDNESS
A measure of the degree of a material's resistance to indentation. It is usually determined by measuring resistance to penetration, by such tests as Brinell, Rockwell, and Vickers.

HEAT ANALYSIS
Formerly known as ladle analysis.

HEAT-EXCHANGER TUBE
A tube for use in apparatus in which fluid inside the tube will be heated or cooled by fluid outside the tube. The term usually is not applied to coiled tubes or to tubes for use in refrigerators or radiators.

HEAT TREATMENT OF STEEL
A combination of heating and cooling operations applied to a metal or alloy in the solid state to obtain desired conditions or properties. Heating for the sole purpose of hot working is excluded from the meaning of this definition. See various types below:

Age hardening An aging process that increases hardness and strength. Ordinarily ductility decreases. Age hardening usually follows rapid cooling or cold working. Hardening is a result of a precipitation process, often sub-microscopic, which occurs when a supersaturated solid solution is naturally aged at atmospheric temperature or artificially aged in some specific range of elevated temperature. Aging occurs more rapidly at higher temperatures. (Synonymous with precipitation hardening).

Air hardening Heating a suitable grade of steel with high hardenability above the critical temperature range and then cooling in air for the purpose of hardening.

Anneal — The annealing process is a combination of a heating cycle, a holding period and a controlled cooling cycle. Annealing is used to obtain a variety of results, among which are: to soften or alter the grain structure of steel, to develop formability, machinability, and required mechanical properties, or to relieve residual stresses. The temperatures and cooling rates used depend on which results are desired. It is generally desirable to use more specific terms in describing the anneal.

Bright Anneal — Carried out in a controlled furnace atmosphere, so that surface oxidation is reduced to a minimum and the tube surface remains relatively bright:

Dead Soft Anneal — A heat treatment applied to achieve maximum softness and ductility.

Finish Anneal — See stress relief anneal.

Full Anneal — Heating to a temperature above the upper critical (above 1650 F.) and ***slow*** cooling below the lower critical, usually in a furnace.

Soft Anneal — When maximum softness and ductility are required without change in grain structure, tubing should be ordered soft annealed. This process consists of heating to a temperature slightly below the critical temperature and cooling in still air. Usually performed in 1250 / 1350 F. range for carbon steel.

Solution Anneal — Heating steel into a temperature range wherein certain elements or compounds dissolve, followed by cooling at a rate sufficient to maintain these elements in solution at room temperature. The expression is normally applied to stainless and other special steels.

Spheroidizing Anneal — A general term which refers to heat treatments that promote spheroidal or globular forms of carbide in carbon or alloy steels.

Stabilizing Anneal — A treatment applied to austenitic stainless steels wherein carbides of various forms are deliberately precipitated. Sufficient additional time is provided at the elevated temperature to diffuse chromium into the areas adjacent to the carbides (usually grain boundaries). This treatment is intended to lessen the chance of intergranular corrosion.

Stress Relief Anneal — Often referred to as "finish annealing" involves heating to a suitable temperature, holding long enough to reduce residual stresses and then cooling slowly enough to minimize the development of new residual stresses. Stress relieving normally takes place in the 950 F - 1150 F temperature range for carbon steels.

Normalize — Normalizing is a process which consists of heating to a temperature approximately 100 F. above the upper critical temperature (above 1650 F.) and cooling in still air at room temperature. Normalizing is utilized to obtain moderate increases in hardness in medium carbon steels or to alter the weld microstructure in lower carbon steels.

Quenching — A process of rapid cooling from an elevated temperature, by contact with liquids or gases. Quenched hardenable steels usually are extremely brittle and are not suitable for use unless subsequently tempered.

Tempering — Reheating quenched or normalized steel to a temperature below the transformation range (lower critical) followed by any desired rate of cooling. Tempering reduces brittleness and develops the desired hardness, structure and properties.

HEAT TREATMENT OF ALUMINUM

Aluminum alloys are divided into two distinct groups based upon their reaction to thermal treatment. Heat-treatable tubing alloys (2000, 6000 and 7000 series) can be strengthened by thermal treatment and subjected to repeated heat treatment cycles without harmful effects. Non-heat-treatable tubing alloys (1000, 3000 and 5000 series) can be strengthened only by cold working. Applicable terms are listed below:

Aging — Precipitation from solid solution resulting in a change in properties of an alloy, usually occuring slowly at room temperature (natural aging) and more rapidly at elevated temperatures (artificial aging).

Age Hardening — An aging process which results in increased strength and hardness.

Age Softening — The loss of strength and hardness at room temperature which takes place in certain alloys due to the spontaneous reduction of residual stresses in the strain hardened structure.

Annealing — A thermal treatment to soften metal by removal of stress resulting from cold working or by coalescing precipitates from solid solution. Performed on all aluminum alloys at approx 650 F. soaking temperature.

Homogenizing — A high temperature soaking treatment to eliminate or reduce segregation by diffusion. Performed on the ingot at 900 - 1000 F. temperature range.

Precipitation Hardening — See "Aging" — The final step in solution heat treating. Artificial aging is performed at temperatures ranging from 240 - 460 F. depending upon heat-treatable aluminum alloy.

Solution Heat Treating — Heating an alloy at a suitable temperature for sufficient time to allow soluble constituents to enter into solid solution where they are retained in a super-saturated state after quenching. Performed in the temperature range of 825 - 980 F. depending upon heat-treatable aluminum alloy and followed by aging.

Stabilizing — A thermal treatment to reduce internal stresses in order to promote dimensional and mechanical property stability. (See "Age Softening" and "Stress Relieving"). Applied only to 3000 and 5000 series alloys in cold worked temper. Performed at about 350° F.

Stress Relieving — The reduction of the effects of internal residual stresses by thermal or mechanical means.

HOT FINISHED SEAMLESS TUBING
Tubing produced by rotary piercing, extrusion, and other hot working processes without subsequent cold finishing operations.

HOT ROLLED ERW TUBING
As welded electric resistance welded tubing made from hot rolled strip, sheet or bands.

HOT SHORTNESS (RED SHORTNESS)
A condition encountered in some metals wherein ductility is lessened at hot working temperatures.

HOT WORKING
The mechanical working of metal above the recrystallization temperature.

HUEY TEST
A corrosion test for stainless steels. The weight loss per unit area is measured after each of five 48-hour boils in 65% nitric acid. The test results are calculated to and reported as the average corrosive rate of the five oils in inches per month (ipm) corrosion rates. The test is used to determine the suitability of a material for nitric acid service. Since most of the weight loss

is due to intergranular attack, the Huey test can be used as an indication of the resistance of a stainless steel to intergranular corrosion.

HYDROSTATIC TEST
There are several methods of determining the toughness of a steel, but the Izod and Charpy notched-bar tests are used quite widely. In both tests, the samples are cooled or heated to the desired test temperature, then struck once with a pendulum which fractures the specimen. The energy required to fracture the specimen, the impact strength, is measured in foot-pounds.

INCLUSIONS
Particles of nonmetallic impurities, usually oxides, sulphides, silicates, which are mechanically held in metals and alloys during solidification.

INDUCTION HEATING
A process of heating by electrical induction.

INGOT
A cast metal shape suitable for subsequent rolling or forging.

INGOT MOLD
A mold in which ingots are cast. Molds may be circular, square, or rectangular in shape, with walls of various thickness. Some molds are of larger cross section at the bottom; others are larger at the top.

INTEGRAL FINNED TUBING
Tubing with raised surface fins formed from the wall of the tube itself.

INTERGRANULAR CORROSION
A type of electrochemical corrosion that progresses preferentially along the grain boundaries of an alloy, usually because the grain boundary regions contain material anodic to the central regions of the grain.

INTERNAL SOUNDNESS
Refers to condition of inside of material — lack of defects, pipe, segregation, non-uniformity of composition.

ISOTHERMAL ANNEAL — See heat treatment.

IZOD IMPACT TEST — See Impact Strength Testing.

JOMINY TEST
Hardenability test performed usually on alloy steels to determine depth and degree of hardness resulting from a standard end quenching method with cold water.

KILLED STEEL
Steel deoxidized with an agent such as silicon or aluminum to reduce the free oxygen content so that no harmful reaction occurs between carbon and oxygen during solidification.

LADLE
A large vessel into which molten steel or molten slag is received and handled.

LADLE ANALYSIS
Chemical analysis obtained from a sample taken during the pouring of the steel, i.e., heat analysis.

LAMINATIONS
Defects resulting from the presence of blisters, seams or foreign inclusions aligned parallel to the worked surface of a metal.

LAP
A surface defect caused from folding the surface of an ingot, bloom or bar during hot rolling operations and then rolling or forging the fold into the surface.

MACHINABILITY
A measure of the relative ease with which steel may be machined.

MACHINING
The deliberate removal of metal by one or more of several processes.

MACROETCH
A testing procedure for locating and identifying porosity, pipes, bursts, unsoundness, inclusions, segregations, carburization, flow lines from hot working, etc. Surface of the test piece should be reasonably smooth or even polished. After applying a suitable etching solution, the structure developed by the action of the reagent may be observed without a microscope.

MAGNAFLUX TEST
This test is conducted by suitable magnetizing the material and applying a prepared wet or dry magnetic powder or fluid which adheres to it along lines of flux leakage. It shows the existence of surface and slightly subsurface non-uniformities.

MALLEABILITY
The property that determines the ease of deforming a metal when the material is subjected to rolling or hammering. The more malleable metals can be hammered or rolled into thin sheet more easily than others.

MANDREL
(1) A device used to retain the cavity in hollow metal products during working. (2) A metal bar around which other metal may be cast, bent, formed or shaped.

MARAGING
A process of improving the mechanical strength of certain ferrous alloys. The name was derived from two hardening reactions: martensite and aging. The maraging strengthening mechanism is based on the age hardening (precipitation hardening) of extra-low carbon martensite.

MARTENSITE
A constituent in quenched steel formed without diffusion and only during rapid cooling below the martensitic start (Ms) temperature. Martensite is the hardest of the transformation products of austenite.

McQUAID-EHN TEST
A special test for revealing the austenitic grain size of ferritic steels when the steel is heated to 1700°F. and carburized. There are eight standard McQuaid-Ehn grain sizes — sizes 5 to 8 are considered fine grain and sizes under 5 are considered coarse grain.

MECHANICAL PROPERTIES
Those properties of a material that reveal the elastic and inelastic reaction when force is applied, or that involve the relationship between stress and strain — for example, the modulus of elasticity, hardness, tensile strength and fatigue limit. These properties have often been referred to as "physical properties", but the term "mechanical properties" is correct.

MECHANICAL TUBING
Used for a variety of mechanical and structural purposes, as opposed to pressure tubing, which is used to contain or conduct fluids or gases under pressure. It may be hot finished or cold drawn. It is commonly manufactured to consumer specifications covering chemical analysis and mechanical properties.

METALLOGRAPHY
The science dealing with the constitution, and structure of metals and alloys as revealed by the unaided eye or by such tools as low powered magnification, optical microscope, electron microscope and diffraction or X-ray techniques.

METRIC SYSTEM OF MEASUREMENTS
In the metric system of measurements, the principal unit for length is the meter; the principal unit for volume, the liter; and the principal unit for weight, the gram. The following prefixes are used for sub-divisions and multiples: milli 1/1000; centi 1/100; deci 1/10; deca 10; hecto 100; kilo 1000. In abbreviations, the sub-divisions are frequently used with a smaller letter and the multiples with a capital letter, although this practice is not universally followed everywhere the metric system is used. All the multiples and the sub-divisions are not used commercially. Those ordinarily used for length are kilometer, meter, centimeter, and milimeter; for area, square meter, square centimeter and square milimeter; for volume, cubic meter, cubic decimeter (liter), cubic centimeter, and cubic milimeter. The most commonly used weights are the kilogram and gram. The metric system was legalized in the United States by an Act of Congress in 1866.

MICROCLEANLINESS
Refers to the extent or quality of nonmetallic inclusions observed by examination under a microscope.

MICRO-ETCH
Micro-etching is used for the examination of a sample under a microscope. Etching solutions tend to reveal structural details because of preferential chemical attack on the polished surface.

MINIMUM WALL
Any wall having tolerances specified all on the plus side.

MODULUS OF ELASTICITY
The ratio of stress applied to a material and the resulting strain occuring at the stresses below the elastic limit.

NITRIDING
A process of case hardening in which a ferrous alloy, usually of special composition, is heated in an atmosphere of cracked ammonia or in contact with nitrogenous material to produce

surface hardening without quenching by the absorption of nitrogen. Nitriding is normally conducted in a range from 900 to 1000 F.

NON-DESTRUCTIVE TESTING
Methods of detecting defects without destroying or permanently changing the material being tested. Test methods include ultrasonic, eddy current, flux leakage, magnetic particle, liquid, penetrant and X-ray.

NOTCH BRITTLENESS
Susceptibility of a material to brittle fracture at points of stress concentration.

NOTCH SENSITIVITY
A measure of the reduction in strength of a metal caused by the presence of stress concentration.

OVALITY
The difference between the maximum and minimum outside diameters of any one cross section of a tube. It is a measure of deviation from roundness.

OXALIC ACID ETCH TEST
A quick metallographic test which is sometimes used to screen stainless steels before intergranular corrosion testing. This test is specified with a referee test such as the Copper-Copper Sulfate or Huey test.

OXIDATION
In its simplest terms, oxidation means the combination of any substance with oxygen. Scale developed during heat treatment is a form of oxidation.

OXIDE
A compound consisting of oxygen and one or more metallic elements.

PASSIVATE
The changing of the chemically active surface of a metal to a much less active state by the application of the proper chemical treatment or by applying an induced electrical current and voltage for cathodic or anodic protection from corrosion. An example of chemically passivating stainless steel would be to immerse stainless in a hot solution of approximately 10 to 20 percent by volume nitric acid and water.

PHOTOMICROGRAPH
A photographic reproduction of an object magnified more than ten times used to show microstructure characteristics of steel.

PHYSICAL PROPERTIES
Those properties not specifically related to reaction to external forces. These include such properties as density, electrical resistance, co-efficient of thermal conductivity.

PICKLING
Use of solutions, usually acids, to remove surface oxides from a tube, may also be used to produce a desired surface finish.

PIERCING
A seamless tubemaking method in which a hot billet is gripped and rotated by rolls or cones and directed over a piercer point which is held on the end of a mandrel bar.

PIT
A sharp, usually small, depression in the surface of metal.

POROSITY
Unsoundness caused in cast metals by the presence of blowholes or shrinkage cavities.

PRESSURE TUBING
Tubing produced for the purpose of containing or conducting fluids or gases under pressure.

PRODUCT ANALYSIS — Formerly known as check analysis.

PROFILOMETER
An instrument used for measuring surface finish. The vertical movements of a stylus as it traverses the surface are amplified electromagnetically and recorded (or indicated) as the surface roughness.

PROOF STRESS
The load per square inch of the original cross-sectional area which, when removed, has caused a permanent elongation not exceeding a defined amount (usually 0.0001" per inch of gage length). A test of this type is more commonly used in Europe than in this country, where it largely has been replaced by yield strength measurements.

PYROMETER
An instrument of any of various types used for measuring temperatures.

QUENCHING — See heat treatment.

RANDOM LENGTH
Tubing produced to a permissible variation in length. (Frequently seven feet.)

RECRYSTALLIZATION
The reversion of distorted cold worked microstructure to a new, stain-free structure during annealing.

REDUCTION OF AREA
A measure of ductility determined in a tensile test. It is the maximum reduction, at the fracture, of the cross section area of a specimen, as compared with its original cross section area.

RIMMED STEEL
A steel which forms a relatively clean outer layer (rim) during solidification. Sheet and strip made from such steel has good surface quality and is frequently used for ERW tubing.

ROCKWELL HARDNESS — See hardness.

ROTO-ROCK (TUBE REDUCING OR ROCKRITE)
A method of cold finishing tubing in which a machine rolls or rocks a split die over a tube. The tube is supported on the inside by a tapered mandrel.

SCALE
An oxide of iron which forms on the surface of hot steel.

SEAM
A tight, but unwelded imperfection on the surface of a wrought metal product.

SEGREGATION
Nonuniform distribution of alloying elements, impurities or microphases.

SEMI-KILLED STEEL
Steel that is incompletely deoxidized to permit the evolution of carbon monoxide, thereby offsetting solidification shrinkage.

SENSITIZATION
Sensitization of stainless steel is defined as a susceptibility to preferential grain boundary attack. Material which exhibits grain boundary carbide precipitation may or may not be sensitized.

SOAK
To hold an ingot, slab, bloom, billet or other piece of steel in a hot furnace, pit or chamber to secure uniform temperature.

SOAKING PIT
A furnace or pit for the heating of ingots of steel to make their temperature uniform prior to rolling or forging.

SOFT ANNEAL — See heat treatment.

SPECIAL SMOOTH I.D. (SSID) — See Electric Resistance Welding Tubing.

SPECIFICATION
A document defining the measurements, tests, and other requirements to which a product must conform — typically covering chemistry, mechanical properties, tolerances, finish, reports, marking and packaging.

SPHEROIDIZE ANNEAL — See heat treatment.

SPINNING
A type of forming (hot or cold) which involves rotating a tube at high speed against fixed or rolling tools for the purpose of altering shape, size, etc.

STABILIZING ANNEAL — See heat treatment.

STRAIN
A measure of the change in size or shape of a body under stress, referred to its original size or shape.

STRAIN HARDENING
Modification of a metal structure by cold working resulting in an increase in strength and hardness with loss of ductility.

STRESS
Force per unit of area measured in pounds per square inch (psi). The three kinds of stresses are tensile, compressive, and shear.

Flexure involves a combination of tensile and compressive stress. Torsion involves shear stress.

STRESS CORROSION CRACKING
Cracking of metals under combined action of temperature, corrosion and stress. The stress can be either applied or residual. Austenitic stainless steels are especially susceptible to cracking in chloride containing and some caustic environments.

STRESS RELIEF ANNEAL — See heat treatment.

STRIP
A flat-rolled steel product which serves as the raw material for welded tubing.

SUNK OR SINK DRAWN
Tubing drawn through a die with no inside mandrel to control I.D. or wall thickness.

SWAGED
A mechanical reduction of the cross sectional area of a metal, performed hot or cold by forging, pressing or hammering.

TAPPING
The act of pouring molten metal from a furnace into a ladle.

TEEMING
Act of pouring molten metal from a ladle into an ingot mold.

TEMPERING — See heat treatment.

TENSILE STRENGTH
The maximum load per square inch of original cross-sectional area carried during a tension test to failure of the specimen. This term is preferred over the formerly-used ultimate strength.

THERMAL CONDUCTIVITY
A measure of the ease with which heat is transmitted through a material.

TOLERANCE
Permissible variation.

TORSION
A twisting action resulting in shear stresses and strains.

TOUGHNESS
A measure of ability to absorb energy and deform plastically before fracturing.

TRANSFORMATION TEMPERATURE
The temperature at which a change in phase occurs in steels. The term is sometimes used to denote the limiting temperature of a transformation range.

TRANSVERSE TENSION TEST
A tension test for evaluating mechanical properties of a material in a direction transverse to that of rolling.

TURNING
A method for removing the surface from a work piece by bringing the cutting edge of a tool against it while the piece or tool is rotated.

ULTIMATE STRENGTH — See tensile strength.

ULTRASONIC TESTING
The method of detecting defects in tubes or welds by passing high frequency sound waves into a material then monitoring and evaluating the reflected signals.

UPSETTING
A metal-working operation similar to forging, generally used to thicken the ends of tubes prior to threading.

VICKERS HARDNESS TEST — See hardness.

WORK HARDENING
Hardness developed in metal as a result of cold working. See cold working.

YIELD POINT
The first stress in a material measured as load per unit of original cross-sectional area at which an increase in strain occurs without an increase in stress.

YIELD STRENGTH
The stress at which a material exhibits a specified deviation from proportionality of stress and strain. An offset of 0.2% is most frequently used.

B

A.N.S.I. Pipe Schedules, Weights and Dimensions

PIPE SIZE	O.D. in Inches	WEIGHTS AND DIMENSIONS OF SEAMLESS AND WELDED PIPE													DBLE. E.H.
		5	10	20	30	40	STD.	60	80	E.H.	100	120	140	160	
1/8	.405	.035 .1383	.049 .1863			.068 .2447	.068 .2447		.095 .3145	.095 .3145					
1/4	.540	.049 .2570	.065 .3297			.088 .4248	.088 .4248		.119 .5351	.119 .5351					
3/8	.675	.049 .3276	.065 .4235			.091 .5676	.091 .5676		.126 .7388	.126 .7388					
1/2	.840	.065 .5383	.083 .6710			.109 .8510	.109 .8510		.147 1.088	.147 1.088				.187 1.304	.294 1.714
3/4	1.050	.065 .6838	.083 .8572			.113 1.131	.113 1.131		.154 1.474	.154 1.474				.218 1.937	.308 2.441
1	1.315	.065 .8678	.109 1.404			.133 1.679	.133 1.679		.179 2.172	.179 2.172				.250 2.844	.358 3.659
1¼	1.660	.065 1.107	.109 1.806			.140 2.273	.140 2.273		.191 2.997	.191 2.997				.250 3.765	.382 5.214
1½	1.900	.065 1.274	.109 2.085			.145 2.718	.145 2.718		.200 3.631	.200 3.631				.281 4.859	.400 6.408
2	2.375	.065 1.604	.109 2.638			.154 3.653	.154 3.653		.218 5.022	.218 5.022				.343 7.444	.436 9.029
2½	2.875	.083 2.475	.120 3.531			.203 5.793	.203 5.793		.276 7.661	.276 7.661				.375 10.01	.552 13.70
3	3.500	.083 3.029	.120 4.332			.216 7.576	.216 7.576		.300 10.25	.300 10.25				.437 14.32	.600 18.58
3½	4.000	.083 3.472	.120 4.973			.226 9.109	.226 9.109		.318 12.51	.318 12.51					.636 22.85
4	4.500	.083 3.915	.120 5.613			.237 10.79	.237 10.79	.281 12.66	.337 14.98	.337 14.98		.437 19.01		.531 22.51	.674 27.54
4½	5.000						.247 12.53			.355 17.61					.710 32.53

Top figures: Wall thickness in inches Bottom figures: Weight per foot in pounds

PIPE SIZE	O.D. in Inches	WEIGHTS AND DIMENSIONS OF SEAMLESS AND WELDED PIPE													DBLE. E.H.
		5	10	20	30	40	STD.	60	80	E.H.	100	120	140	160	
5	5.563	.109 6.349	.134 7.770			.258 14.62	.258 14.62		.375 20.78	.375 20.78		.500 27.04		.625 32.96	.750 38.55
6	6.625	.109 7.585	.134 9.289			.280 18.97	.280 18.97		.432 28.57	.432 28.57		.562 36.39		.718 45.30	.864 53.16
7	7.625						.301 23.57			.500 38.05					.875 63.08
8	8.625	.109 9.914	.148 13.40	.250 22.36	.277 24.70	.322 28.55	.322 28.55	.406 35.64	.500 43.39	.500 43.39	.593 50.87	.718 60.63	.812 67.76	.906 74.69	.875 72.42
9	9.625						.342 33.90			.500 48.72					
10	10.750	.134 15.19	.165 18.70	.250 28.04	.307 34.24	.365 40.48	.365 40.48	.500 54.74	.593 64.33	.500 54.74	.718 76.93	.843 89.20	1.000 104.1	1.125 115.7	1.000 104.1
11	11.750						.375 45.55			.500 60.07					
12	12.750	.165 22.18	.180 24.20	.250 33.38	.330 43.77	.406 53.53	.375 49.56	.562 73.16	.687 88.51	.500 65.42	.843 107.2	1.000 125.5	1.125 139.7	1.312 160.3	1.000 104.1
14	14.000		.250 36.71	.312 45.68	.375 54.57	.437 63.37	.375 54.57	.593 84.91	.750 106.1	.500 72.09	.937 130.7	1.093 150.7	1.250 170.2	1.406 189.1	
16	16.000		.250 42.05	.312 52.36	.375 62.58	.500 82.77	.375 62.58	.656 107.5	.843 136.5	.500 82.77	1.031 164.8	1.218 192.3	1.437 223.5	1.593 245.1	
18	18.000		.250 47.39	.312 59.03	.437 82.06	.562 104.8	.375 70.59	.750 138.2	.937 170.8	.500 93.45	1.156 208.0	1.375 244.1	1.562 274.2	1.781 308.5	
20	20.000		.250 52.73	.375 78.60	.500 104.1	.593 122.9	.375 78.60	.812 166.4	1.031 208.9	.500 104.1	1.280 256.1	1.500 296.4	1.750 341.1	1.968 379.0	
22	22.000		.250 58.07	.375 86.81	.500 114.8		.375 86.81	.875 197.4	1.125 250.8	.500 114.8	1.375 302.9	1.625 353.6	1.875 403.0	2.125 451.7	
24	24.000		.250 63.41	.375 94.62	.562 140.8	.687 171.2	.375 94.62	.968 238.1	1.218 296.4	.500 125.5	1.531 367.4	1.812 429.4	2.062 483.1	2.343 541.9	
26	26.000		.312 85.60	.500 136.2			.375 102.6			.500 136.2					
28	28.000		.312 92.26	.500 146.8	.625 182.7		.375 110.6			.500 146.8					
30	30.000		.312 98.95	.500 157.5	.625 196.01		.375 118.6			.500 157.5					
32	32.000		.312 105.6	.500 168.2	.625 209.4		.375 126.7			.500 168.2					
34	34.000		.312 112.3	.500 178.9	.625 222.8	.688 244.8	.375 134.7			.500 178.9					
36	36.000		.312 118.9	.500 189.5	.625 236.1	.750 252.3	.375 142.7			.500 189.6					
38	38.000						.375 150.7			.500 200.2					
40	40.000						.375 158.7			.500 210.9					
42	42.000						.375 166.7			.500 221.6					
44	44.000						.375 174.7			.500 232.3					
46	46.000						.375 182.7			.500 243.0					
48	48.000						.375 190.7			.500 253.7					
54	54.000						.375 214.8			.500 285.7					
60	60.000						.375 238.8			.500 317.7					

Top figures: Wall thickness in inches   Bottom figures: Weight per foot in pounds

Ⓒ

Aluminum Pipe Schedules, Weights and Dimensions

NOMINAL PIPE SIZE Inches	SCHEDULE NUMBER ②	OUTSIDE DIAMETER Inches			INSIDE DIAMETER Inches	WALL THICKNESS Inches			WEIGHT PER FOOT Pound	
		Nom	Min ③⑦	Max ①⑦	Nom	Nom.	Min ③	Max ①	Nom ④	Max ③④
Col. 1	Col. 2	Col. 3	Col. 4	Col. 5	Col. 6	Col. 7	Col. 8	Col. 9	Col. 10	Col. 11
1/8	40 ⑤	.405	.374	.420	.269	.068	.060	....	.085	.091
	80 ⑥	.405	.374	.420	.215	.095	.083	....	.109	.118
1/4	40 ⑤	.540	.509	.555	.364	.088	.077	....	.147	.159
	80 ⑥	.540	.509	.555	.302	.119	.104	....	.185	.200
3/8	40 ⑤	.675	.644	.690	.493	.091	.080	....	.196	.212
	80 ⑥	.675	.644	.690	.423	.126	.110	....	.256	.276
1/2	5	.840	.809	.855	.710	.065	.053	.077	.186	....
	10	.840	.809	.855	.674	.083	.071	.095	.232	....
	40 ⑤	.840	.809	.855	.622	.109	.095	....	.294	.318
	80 ⑥	.840	.809	.855	.546	.147	.129	....	.376	.406
	160	.840	.809	.855	ϕ .464	ϕ .188	.164	....	ϕ .453	ϕ .489
3/4	5	1.050	1.019	1.065	.920	.065	.053	.077	.237	....
	10	1.050	1.019	1.065	.884	.083	.071	.095	.297	....
	40 ⑤	1.050	1.019	1.065	.824	.113	.099	....	.391	.422
	80 ⑥	1.050	1.019	1.065	.742	.154	.135	....	.510	.551
	160	1.050	1.019	1.065	ϕ .612	ϕ .219	ϕ .192	....	ϕ .672	ϕ .726
1	5	1.315	1.284	1.330	1.185	.065	.053	.077	.300	....
	10	1.315	1.284	1.330	1.097	.109	.095	.123	.486	....
	40 ⑤	1.315	1.284	1.330	1.049	.133	.116	....	.581	.627
	80 ⑥	1.315	1.284	1.330	.957	.179	.157	....	.751	.811
	160	1.315	1.284	1.330	.815	.250	.219	....	.984	1.062
1-1/4	5	1.660	1.629	1.675	1.530	.065	.053	.077	.383	....
	10	1.660	1.629	1.675	1.442	.109	.095	.123	.625	....
	40 ⑤	1.660	1.629	1.675	1.380	.140	.122	....	.786	.849
	80 ⑥	1.660	1.629	1.675	1.278	.191	.167	....	1.037	1.120
	160	1.660	1.629	1.675	1.160	.250	.219	....	1.302	1.407

① Product size ranges listed in this table do not necessarily indicate availability.

② In accordance with ANSI B36.10 and B36.19.

ϕ ③ Based on standard tolerances for pipe on page 6-4-11.

④ Based on density of 0.098 lb. per cu. in., the density of 6061 alloy, and on nominal dimensions and plain ends. For 3003 alloy multiply by 1.01. For 6063 alloy multiply by 0.99.

⑤ Also designated as "Standard" Pipe.

⑥ Also designated as "Extra Heavy" or "Extra Strong" Pipe.

⑦ For schedules 5 and 10 these values apply to mean outside diameters.

NOMINAL PIPE SIZE Inches	SCHEDULE NUMBER (2)	OUTSIDE DIAMETER Inches			INSIDE DIAMETER Inches	WALL THICKNESS Inches			WEIGHT PER FOOT Pounds	
		Nom	Min (3) (7)	Max (3) (7)	Nom	Nom	Min (3)	Max (3)	Nom (4)	Max (3) (4)
Col. 1	Col. 2	Col. 3	Col. 4	Col. 5	Col. 6	Col. 7	Col. 8	Col. 9	Col. 10	Col. 11
1-1/2	5	1.900	1.869	1.915	1.770	.065	.053	.077	.441	....
	10	1.900	1.869	1.915	1.682	.109	.095	.123	.721	....
	40 (5)	1.900	1.869	1.915	1.610	.145	.127	....	.940	1.015
	80 (6)	1.900	1.869	1.915	1.500	.200	.175	....	1.256	1.357
	160	1.900	1.869	1.915	1.338	.281	.246	....	1.681	1.815
2	5	2.375	2.344	2.406	2.245	.065	.053	.077	.555	....
	10	2.375	2.344	2.406	2.157	.109	.095	.123	.913	....
	40 (5)	2.375	2.351	2.399	2.067	.154	.135	....	1.264	1.365
	80 (6)	2.375	2.351	2.399	1.939	.218	.191	....	1.737	1.876
	160	2.375	2.351	2.399	1.687	.344	.301	....	2.581	2.788
2-1/2	5	2.875	2.844	2.906	2.709	.083	.071	.095	.856	....
	10	2.875	2.844	2.906	2.635	.120	.105	.135	1.221	....
	40 (5)	2.875	2.846	2.904	2.469	.203	.178	....	2.004	2.164
	80 (6)	2.875	2.846	2.904	2.323	.276	.242	....	2.650	2.862
	160	2.875	2.846	2.904	2.125	.375	.328	....	3.464	3.741
3	5	3.500	3.469	3.531	3.334	.083	.071	.095	1.048	....
	10	3.500	3.469	3.531	3.260	.120	.105	.135	1.498	....
	40 (5)	3.500	3.465	3.535	3.068	.216	.189	....	2.621	2.830
	80 (6)	3.500	3.465	3.535	2.900	.300	.262	....	3.547	3.830
	160	3.500	3.465	3.535	2.624	.438	.383	....	4.955	5.351
3-1/2	5	4.000	3.969	4.031	3.834	.083	.071	.095	1.201	....
	10	4.000	3.969	4.031	3.760	.120	.105	.135	1.720	....
	40 (5)	4.000	3.960	4.040	3.548	.226	.198	....	3.151	3.403
	80 (6)	4.000	3.960	4.040	3.364	.318	.278	....	4.326	4.672
4	5	4.500	4.469	4.531	4.334	.083	.071	.095	1.354	....
	10	4.500	4.469	4.531	4.260	.120	.105	.135	1.942	....
	40 (5)	4.500	4.455	4.545	4.026	.237	.207	....	3.733	4.031
	80 (6)	4.500	4.455	4.545	3.826	.337	.295	....	5.183	5.598
	120	4.500	4.455	4.545	ϕ 3.624	ϕ .438	ϕ .383	....	ϕ 6.573	ϕ 7.099
	160	4.500	4.455	4.545	3.438	.531	.465	....	7.786	8.409
5	5	5.563	5.532	5.625	5.345	.109	.095	.123	2.196	....
	10	5.563	5.532	5.625	5.295	.134	.117	.151	2.688	....
	40 (5)	5.563	5.507	5.619	5.047	.258	.226	....	5.057	5.461
	80 (6)	5.563	5.507	5.619	4.813	.375	.328	....	7.188	7.763
	120	5.563	5.507	5.619	4.563	.500	.438	....	9.353	10.10
	160	5.563	5.507	5.619	4.313	.625	.547	....	11.40	12.31
6	5	6.625	6.594	6.687	6.407	.109	.095	.123	2.624	....
	10	6.625	6.594	6.687	6.357	.134	.117	.151	3.213	....
	40 (5)	6.625	6.559	6.691	6.065	.280	.245	....	6.564	7.089
	80 (6)	6.625	6.559	6.691	5.761	.432	.378	....	9.884	10.67
	120	6.625	6.559	6.691	5.501	.562	.492	....	12.59	13.60
	160	6.625	6.559	6.691	ϕ 5.187	ϕ .719	ϕ .629	....	ϕ15.69	ϕ16.94
8	5	8.625	8.594	8.718	8.407	.109	.095	.123	3.429	....
	10	8.625	8.594	8.718	8.329	.148	.130	.166	4.635	....
	20	8.625	8.539	8.711	8.125	.250	.219	....	7.735	8.354
	30	8.625	8.539	8.711	8.071	.277	.242	....	8.543	9.227
	40 (5)	8.625	8.539	8.711	7.981	.322	.282	....	9.878	10.67
	60	8.625	8.539	8.711	7.813	.406	.355	....	12.33	13.31
	80 (6)	8.625	8.539	8.711	7.625	.500	.438	....	15.01	16.21
	100	8.625	8.539	8.711	ϕ 7.437	ϕ .594	ϕ .520	....	ϕ17.62	ϕ19.03
	120	8.625	8.539	8.711	ϕ 7.187	ϕ .719	ϕ .629	....	ϕ21.00	ϕ22.68
	140	8.625	8.539	8.711	7.001	.812	.710	....	23.44	25.31
	160	8.625	8.539	8.711	6.813	.906	.793	....	25.84	27.90
10	5	10.750	10.719	10.843	10.482	.134	.117	.151	5.256	....
	10	10.750	10.719	10.843	10.420	.165	.144	.186	6.453	....
	20	10.750	10.642	10.858	10.250	.250	.219	....	9.698	10.47
	30	10.750	10.642	10.858	10.136	.307	.269	....	11.84	12.79
	40 (5)	10.750	10.642	10.858	10.020	.365	.319	....	14.00	15.12
	60	10.750	10.642	10.858	9.750	.500	.438	....	18.93	20.45
	80 (6)	10.750	10.642	10.858	ϕ 9.562	ϕ .594	ϕ .520	....	ϕ22.29	ϕ24.07
	100	10.750	10.642	10.858	ϕ 9.312	ϕ .719	ϕ .629	....	ϕ26.65	ϕ28.78
12	5	12.750	12.719	12.843	12.438	.156	.136	.176	7.258	....
	10	12.750	12.719	12.843	12.390	.180	.158	.202	8.359	....
	20	12.750	12.622	12.878	12.250	.250	.219	....	11.55	12.47
	30	12.750	12.622	12.878	12.090	.330	.289	....	15.14	16.35
	40	12.750	12.622	12.878	11.938	.406	.355	....	18.52	20.00
	60	12.750	12.622	12.878	11.626	.562	.492	....	25.31	27.33
	80 (6)	12.750	12.622	12.878	ϕ11.374	ϕ .688	ϕ .602	....	ϕ30.66	ϕ33.11

(1) Product size ranges listed in this table do not necessarily indicate availability.

(2) In accordance with ANSI B36.10 and B36.19.

(3) Based on standard tolerances for pipe on page 6-4-11.

(4) Based on density of 0.098 lb. per cu. in., the density of 6061 alloy, and on nominal dimensions and plain ends. For 3003 alloy multiply by 1.01. For 6063 alloy multiply by 0.99.

(5) Also designated as "Standard" Pipe.

(6) Also designated as "Extra Heavy" or Extra Strong" Pipe.

(7) For schedules 5 and 10 these values apply to mean outside diameters.

D

Copper and Brass Pipe Sizes, Weights and Dimensions

Pipe Size	Weights and Dimensions—Iron Pipe Sizes						
	Outside Diam. for All Wgts.	Standard			Extra Strong		
		Inside Diam.	Weight Lbs. per Ft.		Inside Diam.	Weight Lbs. per Ft.	
			Brass	Copper		Brass	Copper
1/8	0.405	0.281	0.253	0.259	0.205	0.363	0.371
1/4	0.540	0.376	0.447	0.457	0.294	0.611	0.625
3/8	0.675	0.495	0.627	0.641	0.421	0.829	0.847
1/2	0.840	0.626	0.934	0.955	0.542	1.23	1.25
3/4	1.050	0.822	1.27	1.30	0.736	1.67	1.71
1	1.315	1.063	1.78	1.82	0.951	2.46	2.51
1-1/4	1.660	1.368	2.63	2.69	1.272	3.39	3.46
1-1/2	1.900	1.600	3.13	3.20	1.494	4.10	4.19
2	2.375	2.063	4.12	4.22	1.933	5.67	5.80
2-1/2	2.875	2.501	5.99	6.12	2.315	8.66	8.85
3	3.500	3.062	8.56	8.75	2.892	11.6	11.8
3-1/2	4.000	3.500	11.2	11.4	3.358	14.1	14.4
4	4.500	4.000	12.7	12.9	3.818	16.9	17.3
5	5.562	5.062	15.8	16.2	4.812	23.2	23.7
6	6.625	6.125	19.0	19.4	5.751	32.2	32.9
8	8.625	8.001	30.9	31.6	7.625	48.4	49.5
10	10.750	10.020	45.2	46.2	9.750	61.1	62.4
12	12.750	12.000	55.3	56.5			

E

Copper Tube Sizes, Weights and Dimensions

TYPE K

Size, inches	Nominal Dimensions, inches			Calculated Values, Based on Nominal Dimensions			
	Outside Diameter	Inside Diameter	Wall Thickness	Cross Sectional Area of Bore, sq inches	External Surface, sq ft per lin ft	Internal Surface, sq ft per lin ft	Weight, pounds per lin ft
¼	.375	.305	.035	.073	.098	.080	0.145
⅜	.500	.402	.049	.127	.131	.105	0.269
½	.625	.527	.049	.218	.164	.138	0.344
⅝	.750	.652	.049	.334	.196	.171	0.418
¾	.875	.745	.065	.436	.229	.195	0.641
1	1.125	.995	.065	.778	.294	.261	0.839
1¼	1.375	1.245	.065	1.22	.360	.326	1.04
1½	1.625	1.481	.072	1.72	.425	.388	1.36
2	2.125	1.959	.083	3.01	.556	.513	2.06
2½	2.625	2.435	.095	4.66	.687	.638	2.93
3	3.125	2.907	.109	6.64	.818	.761	4.00
3½	3.625	3.385	.120	9.00	.949	.886	5.12
4	4.125	3.857	.134	11.7	1.08	1.01	6.51
5	5.125	4.805	.160	18.1	1.34	1.26	9.67
6	6.125	5.741	.192	25.9	1.60	1.50	13.9
8	8.125	7.583	.271	45.2	2.13	1.98	25.9
10	10.125	9.449	.338	70.1	2.65	2.47	40.3
12	12.125	11.315	.405	101.	3.17	2.96	57.8

TYPE L

¼	.375	.315	.030	.078	.098	.082	0.126
⅜	.500	.430	.035	.145	.131	.113	0.198
½	.625	.545	.040	.233	.164	.143	0.285
⅝	.750	.666	.042	.348	.196	.174	0.362
¾	.875	.785	.045	.484	.229	.206	0.455
1	1.125	1.025	.050	.825	.294	.268	0.655
1¼	1.375	1.265	.055	1.26	.360	.331	0.884
1½	1.625	1.505	.060	1.78	.425	.394	1.14
2	2.125	1.985	.070	3.09	.556	.520	1.75
2½	2.625	2.465	.080	4.77	.687	.645	2.48
3	3.125	2.945	.090	6.81	.818	.771	3.33
3½	3.625	3.425	.100	9.21	.949	.897	4.29
4	4.125	3.905	.110	12.0	1.08	1.02	5.38
5	5.125	4.875	.125	18.7	1.34	1.28	7.61
6	6.125	5.845	.140	26.8	1.60	1.53	10.2
8	8.125	7.725	.200	46.9	2.13	2.02	19.3
10	10.125	9.625	.250	72.8	2.65	2.53	30.1
12	12.125	11.565	.280	105.	3.17	3.03	40.4

TYPE M

Size, inches	Nominal Dimensions, inches			Calculated Values, Based on Nominal Dimensions			
	Outside Diameter	Inside Diameter	Wall Thickness	Cross Sectional Area of Bore, sq inches	External Surface, sq ft per lin ft	Internal Surface, sq ft per lin ft	Weight, pounds per lin ft
⅜	.500	450	.025	.159	.131	.118	0.145
½	.625	.569	.028	.254	.164	.149	0.204
¾	.875	.811	.032	.517	.229	.212	0.328
1	1.125	1.055	.035	.874	.294	.276	0.465
1¼	1.375	1.291	.042	1.31	.360	.338	0.682
1½	1.625	1.527	.049	1.83	.425	.400	0.940
2	2.125	2.009	.058	3.17	.556	.526	1.46
2½	2.625	2.495	.065	4.89	.687	.653	2.03
3	3.125	2.981	.072	6.98	.818	.780	2.68
3½	3.625	3.459	.083	9.40	.949	.906	3.58
4	4.125	3.935	.095	12.2	1.08	1.03	4.66
5	5.125	4.907	.109	18.9	1.34	1.28	6.66
6	6.125	5.881	.122	27.2	1.60	1.54	8.92
8	8.125	7.785	.170	47.6	2.13	2.04	16.5
10	10.125	9.701	.212	73.9	2.65	2.54	25.6
12	12.125	11.617	.254	106.	3.17	3.04	36.7

TYPE DWV

1¼	1.375	1.295	.040	1.32	.360	.339	.65
1½	1.625	1.541	.042	1.87	.425	.403	.81
2	2.125	2.041	.042	3.27	.556	.534	1.07
3	3.125	3.030	.045	7.21	.818	.793	1.69
4	4.125	4.009	.058	12.6	1.08	1.05	2.87
5	5.125	4.981	.072	19.5	1.34	1.30	4.43
6	6.125	5.959	.083	27.9	1.60	1.56	6.10
8	8.125	7.907	.109	49.1	2.13	2.07	10.6

F

Round Hollow Structural Tube Sizes, Weights and Dimensions

Actual O.D. (inches)	Wall Thickness (inches)	Weight (pounds/ foot)	Area (square inches) A	Moment of Inertia (inches^4) I	Section Modulus (inches^3) S	Radius of Gyration (inches) r
1.050	.1000	1.01	.2985	.0340	.0648	.3377
	.1250	1.23	.3632	.0396	.0754	.3300
1.315	.1000	1.30	.3817	.0709	.1079	.4310
	.1250	1.59	.4673	.0836	.1272	.4231
1.660	.1000	1.67	.4901	.1497	.1804	.5527
	.1250	2.05	.6028	.1787	.2153	.5445
1.900	.1100	2.10	.6186	.2487	.2618	.6341
	.1250	2.37	.6970	.2759	.2904	.6291
	.1500	2.80	.8247	.3180	.3348	.6210
	.1875	3.43	1.0087	.3742	.3939	.6091
2.375	.1250	3.00	.8836	.5609	.4723	.7967
	.1650	3.89	1.1456	.7033	.5923	.7835
	.2100	4.86	1.4283	.8448	.7114	.7691
	.2500	5.67	1.6690	.9551	.8043	.7565
2.875	.1500	4.36	1.2841	1.1956	.8317	.9649
	.2100	5.98	1.7582	1.5707	1.0926	.9452
	.2500	7.01	2.0617	1.7920	1.2466	.9323
3.500	.1500	5.37	1.5787	2.2191	1.2681	1.1856
	.2100	7.38	2.1705	2.9489	1.6851	1.1656
	.2500	8.68	2.5525	3.3903	1.9373	1.1525
4.000	.1500	6.17	1.8143	3.3668	1.6834	1.3622
	.2100	8.50	2.5004	4.5035	2.2518	1.3421
	.2500	10.01	2.9452	5.2005	2.6002	1.3288
4.500	.1500	6.97	2.0499	4.8547	2.1576	1.5389
	.2100	9.62	2.8303	6.5270	2.9009	1.5186
	.2500	11.35	3.3379	7.5629	3.3613	1.5052
5.562	.2580	14.62	4.2995	15.1588	5.4503	1.8777
	.3750	20.78	6.1114	20.6658	7.4304	1.8389
6.625	.1250	8.68	2.5525	13.4863	4.0713	2.2986
	.1880	12.92	3.8018	19.7089	5.9499	2.2769
	.2800	18.97	5.5814	28.1437	8.4962	2.2455
	.4320	28.57	8.4050	40.4929	12.2243	2.1949

Actual O.D. (inches)	Wall Thickness (inches)	Weight (pounds/ foot)	Area (square inches) A	Moment of Inertia (inches4) I	Section Modulus (inches3) S	Radius of Gyration (inches) r
8.625	.1250	11.35	3.3379	30.1539	6.9922	3.0056
	.1880	16.94	4.9831	44.3632	10.2871	2.9838
	.2500	22.36	6.5777	57.7250	13.3855	2.9624
	.2770	24.70	7.2646	63.3562	14.6913	2.9532
	.3220	28.55	8.3993	72.4931	16.8100	2.9378
	.5000	43.39	12.7627	105.7218	24.5152	2.8781
10.750	.2190	24.63	7.2454	100.4905	18.6959	3.7242
	.2500	28.04	8.2467	113.7200	21.1572	3.7135
	.2790	31.20	9.1779	125.8811	23.4197	3.7035
	.3070	34.24	10.0720	137.4274	25.5679	3.6939
	.3650	40.48	11.9083	160.7431	29.9057	3.6740
	.5000	54.74	16.1007	211.9614	39.4346	3.6283
12.750	.1880	25.22	7.4194	146.3914	22.9633	4.4420
	.2500	33.38	9.8175	191.8345	30.0917	4.4204
	.3300	43.77	12.8762	248.4671	38.9752	4.3928
	.3750	49.56	14.5790	279.3499	43.8196	4.3773
	.4060	53.53	15.7446	300.2249	47.0941	4.3667
	.5000	65.42	19.2423	361.5630	56.7158	4.3347
14.000	.2500	36.71	10.7992	255.3140	36.4734	4.8623
	.3750	54.57	16.0516	372.7800	53.2543	4.8191
	.5000	72.09	21.2058	483.7817	69.1117	4.7764
16.000	.2500	42.05	12.3700	383.6843	47.9605	5.5693
	.3750	62.58	18.4078	562.1140	70.2643	5.5260
	.5000	82.77	24.3474	731.9810	91.4976	5.4831

Square Hollow Structural Tube Sizes, Weights and Dimensions

Size (inches)	Wall Thickness (inches)	Weight (pounds/ foot)	Area (inches²)	Moment of Inertia (inches²)	Section Modulus (inches²)	Radius of Gyration (inches²)
1 x 1	.100	1.12	.3312	.0439	.0897	.3642
	.125	1.34	.3934	.0492	.0983	.3535
1¼x1¼	.100	1.50	.4312	.0935	.1497	.4658
	.125	1.88	.5184	.1074	.1719	.4552
1½x1½	.100	1.80	.5312	.1710	.2280	.5674
	.125	2.19	.6434	.1995	.2659	.5568
	.188	2.94	.865	2.26	.301	.511
2 x 2	.110	2.69	.7914	.4574	.4574	.7603
	.125	3.04	.8934	.5079	.5079	.7540
	.150	3.57	1.050	.58	.580	.743
	.1875	4.31	1.2688	.6667	.6667	.7249
	.250	5.22	1.54	.713	.713	.681
2½x2½	.125	3.89	1.140	1.050	.842	.960
	.1875	5.59	1.6438	1.4211	1.1369	.9298
	.250	7.10	2.0890	1.6849	1.3479	.8981
2¾x2¾	.125	4.51	1.325	1.530	1.113	1.074
	.188	6.63	1.956	2.158	1.569	1.050
3 x 3	.125	4.74	1.39	1.89	1.26	1.160
	.1875	6.86	2.0188	2.5977	1.7318	1.1344
	.250	8.80	2.5890	3.1509	2.1006	1.1032
	.312	10.27	3.03	3.39	2.26	1.060

Size (inches)	Wall Thickness (inches)	Weight (pounds/ foot)	Area (inches²)	Moment of Inertia (inches²)	Section Modulus (inches²)	Radius of Gyration (inches²)
	.125	5.84	1.718	3.260	1.863	1.377
	.156	6.88	2.0240	3.7112	2.1207	1.3541
3½x3½	.1875	8.14	2.3938	4.2904	2.4517	1.3388
	.250	10.50	3.0890	5.2844	3.0196	1.3079
	.3125	12.69	3.7329	6.0826	3.4758	1.2765
	.125	6.33	1.915	4.858	2.429	1.593
	.1875	9.31	2.7383	6.4677	3.2338	1.5369
4 x 4	.250	12.02	3.5354	7.9880	3.9940	1.5031
	.3125	14.52	4.2720	9.2031	4.6016	1.4677
	.375	16.84	4.9543	10.152	5.0760	1.4315
	.500	20.88	6.1416	11.2[illegible]	5.6169	1.3524
	.125	7.27	2.135	6.984	3.104	1.787
4½x4½	.188	10.76	3.166	10.067	4.474	1.762
	.250	14.15	4.173	12.839	5.706	1.738
	.1875	11.86	3.4883	13.208	5.2831	1.9458
	.250	15.42	4.5354	16.595	6.6380	1.9128
5 x 5	.3125	18.77	5.5220	19.489	7.7955	1.8786
	.375	21.94	6.4543	21.946	8.7784	1.8440
	.500	27.68	8.1416	25.521	10.208	1.7705
	.125	8.68	2.552	11.224	4.276	2.093
5¼x5¼	.188	12.89	3.802	16.279	6.202	2.068
	.250	17.02	5.007	20.924	7.971	2.044
	.1875	14.41	4.2383	23.496	7.8322	2.3545
	.250	18.82	5.5354	29.845	9.9482	2.3220
6 x 6	.3125	23.02	6.7720	35.465	11.822	2.2884
	.375	27.04	7.9543	40.436	13.479	2.2547
	.450	32.19	9.4	47.20	15.70	2.230
	.1875	16.85	4.9577	37.698	10.771	2.7575
	.250	22.04	6.4817	48.052	13.729	2.7228
7 x 7	.3125	26.99	7.9389	57.306	16.373	2.6867
	.375	31.73	9.3339	65.544	18.727	2.6499
	.500	40.55	11.927	78.913	22.547	2.5722
	.250	25.44	7.4817	73.382	18.346	3.1318
	.3125	31.24	9.1889	88.095	22.024	3.0963
8 x 8	.375	36.83	10.834	101.46	25.366	3.0603
	.450	44.43	13.01	122.00	30.04	3.05
	.500	47.35	13.927	124.08	31.021	2.9849
	.250	32.23	9.4817	147.89	29.578	3.9494
	.3125	39.74	11.689	179.12	35.824	3.9146
10 x 10	.375	47.03	13.834	208.21	41.642	3.8795
	.450	56.57	16.70	249.00	49.90	3.87
	.500	60.95	17.927	259.81	51.962	3.8069
	.250	39.03	11.4817	262.07	43.679	4.7776
	.312	48.17	14.1688	319.65	53.275	4.7497
12 x 12	.375	57.23	16.8339	375.39	62.565	4.7223
	.450	68.90	20.30	445.00	74.10	4.68
	.500	74.54	21.9270	478.16	79.694	4.6698

H

Rectangular Hollow Structural Tube Sizes, Weights and Dimensions

Size inches		Wall inches	Weight per foot pounds W	Area of metal square inches A	Axis X-X			Axis Y-Y		
Axis Y-Y	Axis X-X				Moment of inertia I	Section modulus S	Radius of gyration r	Moment of inertia I	Section modulus S	Radius of gyration r
2	1	.100	1.80	.5312	.2540	.2540	.6915	.0844	.1689	.3987
		.125	2.19	.6434	.2964	.2964	.6787	.0970	.1940	.3883
3	2	.141	4.32	1.2720	1.4972	.9981	1.0849	.7951	.7951	.7906
		.1875	5.59	1.6438	1.8551	1.2367	1.0623	.9758	.9758	.7704
		.250	7.10	2.0890	2.2030	1.4687	1.0269	1.1466	1.1466	.7409
3½	2	.125	4.51	1.325	2.172	1.241	1.280	.891	.891	.820
		.188	6.63	1.951	3.076	1.758	1.256	1.233	1.233	.795
4	1½	.125	4.51	1.325	2.551	1.275	1.387	.519	.692	.626
		.188	6.63	1.956	3.634	1.817	1.363	.704	.939	.600
4	2	.155	5.78	1.7015	3.3477	1.6738	1.4027	1.1230	1.1230	.8124
		.1875	6.86	2.0188	3.8654	1.9327	1.3837	1.2849	1.2849	.7978
		.250	8.80	2.5890	4.6893	2.3447	1.3458	1.5321	1.5321	.7692
4	3	.125	5.84	1.718	4.018	2.009	1.529	2.555	1.703	1.219
		.156	6.88	2.0240	4.5198	2.2599	1.4944	2.8949	1.9299	1.1959
		.1875	8.14	2.3938	5.2291	2.6146	1.4780	3.3404	2.2269	1.1813
		.250	10.50	3.0890	6.4498	3.2249	1.4450	4.0988	2.7326	1.1519
		.3125	12.69	3.7329	7.4338	3.7169	1.4112	4.7000	3.1333	1.1221
5	2	.125	5.84	1.718	5.371	2.148	1.768	1.233	1.233	.847
		.188	8.64	2.547	7.739	3.096	1.743	1.715	1.715	.821
		.250	10.50	3.040	8.140	3.250	1.640	1.870	1.870	.785
5	3	.1875	9.31	2.7383	8.8629	3.5452	1.7991	4.0118	2.6746	1.2104
		.250	12.02	3.5354	10.949	4.3797	1.7598	4.9195	3.2797	1.1796
		.3125	14.52	4.2720	12.612	5.0448	1.7182	5.6255	3.7504	1.1475
		.375	16.84	4.9543	13.907	5.5628	1.6754	6.1552	4.1034	1.1146
		.500	20.88	6.1416	15.355	6.1418	1.5812	6.6839	4.4559	1.0432
6	3	.125	7.27	2.135	10.433	3.444	2.184	3.535	2.357	1.271
		.1875	10.58	3.1133	13.991	4.6637	2.1199	4.7545	3.1697	1.2358
		.250	13.72	4.0354	17.438	5.8128	2.0788	5.8675	3.9116	1.2058
		.3125	16.65	4.8970	20.287	6.7622	2.0353	6.7592	4.5061	1.1748
		.375	19.39	5.7043	22.612	7.5373	1.9910	7.4560	4.9706	1.1433
		.500	24.28	7.1416	25.629	8.5431	1.8944	8.2672	5.5115	1.0759

Size inches Axis Y-Y	Size inches Axis X-X	Wall inches	Weight per foot pounds W	Area of metal square inches A	Axis X-X Moment of inertia I	Axis X-X Section modulus S	Axis X-X Radius of gyration r	Axis Y-Y Moment of inertia I	Axis Y-Y Section modulus S	Axis Y-Y Radius of gyration r
6	4	.1875	11.86	3.4883	17.160	5.7198	2.2179	9.1952	4.5976	1.6236
		.250	15.42	4.5354	21.574	7.1913	2.1810	11.509	5.7544	1.5930
		.3125	18.77	5.5220	25.346	8.4487	2.1424	13.463	6.7313	1.5614
		.375	21.94	6.4543	28.553	9.5178	2.1033	15.097	7.5486	1.5294
		.500	27.68	8.1416	33.213	11.071	2.0198	17.400	8.7002	1.4619
6	4½	.125	8.68	2.553	13.670	4.556	2.309	8.778	3.901	1.850
		.188	12.89	3.802	19.867	6.622	2.284	12.691	5.640	1.825
		.250	17.02	5.007	25.595	8.531	2.261	16.256	7.225	1.802
7	5	.1875	14.41	4.2383	29.380	8.3943	2.6329	17.552	7.0210	2.0350
		.250	18.82	5.5354	37.341	10.669	2.5973	22.241	8.8963	2.0045
		.3125	23.02	6.7720	44.396	12.685	2.5604	26.365	10.546	1.9731
		.375	27.04	7.9543	50.646	14.470	2.5233	29.985	11.994	1.9416
		.500	34.48	10.142	60.642	17.326	2.4453	35.688	14.275	1.8759
8½	2	.125	8.68	2.553	20.467	4.815	2.825	1.981	1.981	.879
		.188	12.89	3.802	29.791	7.008	2.797	2.767	2.767	.853
		.250	17.02	5.007	38.354	9.024	2.770	3.447	3.447	.829
8	4	.1875	14.41	4.2383	34.828	8.7070	2.8666	11.923	5.9614	1.6772
		.250	18.82	5.5354	44.230	11.058	2.8267	15.030	7.5148	1.6478
		.3125	23.02	6.7720	52.533	13.133	2.7852	17.722	8.8610	1.6177
		.375	27.04	7.9543	59.864	14.966	2.7433	20.042	10.021	1.5874
		.500	34.48	10.142	71.475	17.869	2.6548	23.567	11.784	1.5244
8	6	.1875	16.85	4.9577	45.772	11.443	3.0385	29.548	9.8493	2.4413
		.250	22.04	6.4817	58.362	14.590	3.0007	37.608	12.536	2.4088
		.3125	26.99	7.9389	69.617	17.404	2.9613	44.784	14.928	2.3751
		.375	31.73	9.3339	79.643	19.911	2.9211	51.143	17.048	2.3408
		.500	40.55	11.927	95.916	23.979	2.8358	61.374	20.458	2.2684
10	6	.250	25.44	7.4817	100.35	20.070	3.6623	45.879	15.293	2.4763
		.3125	31.24	9.1889	120.45	24.089	3.6205	54.903	18.301	2.4444
		.375	36.83	10.834	138.69	27.739	3.5780	63.026	21.009	2.4119
		.500	47.35	13.927	169.48	33.896	3.4884	76.541	25.514	2.3443
12	8	.250	32.60	9.590	196.00	32.600	4.5200	105.000	26.300	3.3100
		.312	40.25	11.900	239.00	39.800	4.4900	128.000	32.000	3.2800
		.375	47.85	14.100	279.00	46.500	4.4500	149.000	37.300	3.2600
		.450	56.67	16.700	325.00	54.100	4.4100	173.000	43.200	3.2200

†The weight, area, and other properties were calculated on the basis of a section with rounded corners and consequently show the actual section properties rather than the idealized version considering the corners as square.

D

Formulas for Determining Moment of Inertia and Section Modulus

Moment of Inertia $- I = 0.0491\ (OD^4 - ID^4)\ INS^4$

e.g. 4″ S/40 $- I = 0.0491\ (4.5^4 - 4.026^4)\ INS^4$

$= 0.0491\ (410.0625 - 262.7211)\ INS^4$

$= 0.0491\ (147.3414)\ INS^4$

$= 7.2344\ INS^4$

Section Modulus $- Z = \frac{I}{Y} = 0.0982 \left(\frac{OD^4 - ID^4}{OD}\right) INS^3$

$= 0.0982 \left(\frac{410.0625 - 262.7211}{4.5}\right) INS^3$

$= 0.0982 \left(\frac{147.3414}{4.5}\right) INS^3$

$= 0.0982\ (32{,}7425) INS^3$

$= 3.2153\ INS^3$

OD = Outside Diameter
ID = Inside Diameter
I = Moment of Inertia
Y = Distance of farthest Fiber from Neutral Axis $\left(\frac{OD}{2}\right)$
Z = Section Modulus

J

Selection Guide—Commercial or Precision Tooling for Tube Bending

Bending tools are designed and manufactured according to the performance expected in their application. Precision class tooling is held to the highest specifications and closest tolerances in the bending industry. Commercial class tool manufacturing tolerances, die material, and hardening are selected to meet application and production requirements.

The following tables will assist you in selecting the class of tooling best suited to meet your bending requirements.

PRECISION TOOLING is designed for tubing of lighter wall thickness on smaller radii. See below for guidelines between classes. Tooling for the aircraft/aerospace field is usually built to precision class standards.

COMMERCIAL TOOLING is for heavier wall tubing and for other less critical applications.

SELECTING DIES AND TOOL CLASS
Locate each tube diameter across top line of the chart, then read down to the closest wall thickness line. Figure given is the Wall Factor, or tube diameter divided by the wall thickness.

To determine tooling class read down the appropriate tube diameter column to the closest wall thickness line. If this point is BELOW line "A", PRECISION class tools are indicated for bends on 1D radii. Points below lines B through E indicate precision tooling for respective bend radii of 1½D, 2D, 3D, and 4D. Points above each line can usually be accommodated on commercial class tools.

LINE A-1D (Centerline Radius = 1 × Tube O.D.)
LINE B-1½D (Centerline Radius = 1½ × Tube O.D.)
LINE C-2D (Centerline Radius = 2 × Tube O.D.)
LINE D-3D (Centerline Radius = 3 × Tube O.D.)
BENDS ABOVE RADIUS LINE-COMMERCIAL CLASS TOOLS

Wall Thickness (inches)	Tube Outside Diameter—Inches																			
	¼	⅜	½	⅝	¾	⅞	1	1¼	1½	1¾	2	2¼	2½	3	3½	4	4½	5	5½	6
.120								10.4	12.5	14.5	16.6	18.7	20.8	25.0	29.1	33.3	37.5	41.7	45.8	50.0
.109								11.4	13.8	16.0	18.3	20.6	22.9	27.5	32.1	36.6	41.2	45.9	50.4	55.0
.095							10.5	13.1	15.7	18.4	21.0	23.6	26.3	31.5	36.8	42.1	47.3	52.6	57.9	63.2
.083						10.5	12.0	15.0	18.0	21.0	24.0	27.1	30.1	36.1	42.1	48.1	54.2	60.2	66.3	72.3
.065					11.5	13.4	15.3	19.2	23.0	26.9	30.7	34.6	38.4	46.1	53.8	61.5	69.2	76.9	84.6	92.3
.058				10.7	12.9	15.0	17.2	21.5	25.8	30.1	34.4	38.7	43.1	51.7	60.3	69.0	77.6	86.2	94.8	103.4
.049			10.2	12.7	15.3	17.8	20.4	25.5	30.6	35.7	40.8	45.9	51.0	61.2	71.4	81.6	91.8	102.0	112.2	122.4
.042		8.9	11.9	14.9	17.9	20.8	23.8	29.8	35.7	41.7	47.6	53.5	59.5	71.4	83.3	95.2	107.1	119.0	130.9	142.9
.035		10.7	14.2	17.8	21.4	25.0	28.5	35.7	42.8	50.0	57.1	64.3	71.4	85.7	100.0	114.2	128.5	142.9	157.1	171.4
.032		11.7	15.6	19.5	23.4	27.3	31.2	39.1	46.8	54.6	62.5	70.3	78.1	93.8	109.4	125.0	140.6	156.3	171.9	187.5
.028		13.3	17.8	22.3	26.7	31.2	35.7	44.6	53.5	62.5	71.4	80.3	89.2	107.1	125.0	142.8	160.7	178.6	196.4	214.3
.025		15.0	20.0	25.0	30.0	35.0	40.0	50.0	60.0	70.0	80.0	90.0	100.0	120.0	140.0	160.0	180.0	200.0	220.0	240.0
.022		17.0	22.7	28.4	34.1	39.8	45.4	56.8	68.1	79.5	90.9	102.2	113.6	136.3	159.0	181.8	204.5	227.3	250.0	272.7
.020	12.5	18.7	25.0	31.2	37.5	43.7	50.0	62.5	75.0	87.5	100.0	112.5	125.0	150.0	175.0	200.0	225.0	250.0	275.0	300.0
.016	15.6	23.4	31.2	39.0	46.8	54.6	62.5	78.1	93.7	109.3	125.0	140.6	156.2	187.5	218.8					

K

Mandrel and Wiper Die Selection Chart

This chart is your guide to determining when a mandrel or wiper die is generally required, plus the mandrel type and number of balls recommended for high quality, wrinkle-free bends with maximum 2½% flattening.

KEY TO CHART

$$\text{Wall Factor} = \frac{\text{Tube Outside Diameter}}{\text{Wall Thickness}}$$

(Or see Table 4, page 13)

$$\text{Radius to O.D. Factor} = \frac{\text{Bend Centerline Radius}}{\text{Tube Outside Diameter}}$$

Tooling Identification:

P = Plug Mandrel
F = Form Mandrel
M = Ball Mandrel
Figure — Number of Balls on Mandrel

Example: Bend 3″ O.D. Tube with .049″ wall on 6″ *C.L.* Radius

$$\text{Wall Factor} = \frac{3}{.049} = 61.2$$

$$\text{Radius to O.D. Factor} = \frac{6}{3} = 2$$

Read down to closest wall factor (60), across to fourth column (2D). Answer:

```
M   3   W
|   |   └─── Wiper Die
|   └──────Three Ball      OR:
└─────── Mandrel
```

Three ball mandrel with wiper die.

NOTE: 1. Light shaded area at top of table—No wiper die is required when bending brass or copper.
2. Light shaded area, factors 70 to 175—Use Thin Wall type Uni-Flex mandrel.
3. Darker shaded area, factors 200 to 275—Use Ultra-Thin Wall type Uni-Flex mandrel.

		CENTERLINE RADIUS TO O.D. FACTOR ("D" OF BEND)							
	WALL FACTOR	**1**	**1.25**	**1.50**	**2**	**2.50**	**3**	**4**	**5**
	10	F	F	F	F	P	P		
SEE NOTE 1	15	M1W	M1W	M1	M1	P	P		
	20	M2W	M1W	M1W	M1	F	F	P	
	25	M3W	M2W	M1W	M1W	M1	F	F	
	30	M3W	M3W	M2W	M2W	M1W	M1	F	F
	35	M3W	M3W	M3W	M2W	M2W	M2W	M2	M1
	40	M4W	M3W	M3W	M3W	M3W	M3W	M2W	M2
	45	M4W	M3W	M3W	M3W	M3W	M3W	M2W	M2W
	50	M4W	M3W	M3W	M3W	M3W	M3W	M2W	M2W
	60	M4W	M4W	M3W	M3W	M3W	M3W	M2W	M2W
SEE NOTE 2	70	M5W	M5W	M5W	M3W	M3W	M3W	M3W	M2W
	80	M5W	M5W	M5W	M5W	M3W	M3W	M3W	M2W
	90	M5W	M5W	M5W	M5W	M3W	M3W	M3W	M3W
	100	M5W	M5W	M5W	M5W	M5W	M3W	M3W	M3W
	125	M5W	M5W	M5W	M5W	M5W	M5W	M4W	M4W
	150	M6W	M6W	M6W	M6W	M5W	M5W	M4W	M4W
	175	M6W	M7W	M8W	M7W	M7W	M6W	M6W	M6W
SEE NOTE 3	200	M6W	M8W	M10W	M10W	M9W	M9W	M8W	M8W
	225		M9W	M10W	M10W	M10W	M10W	M10W	M10W
	250			M10W	M10W	M10W	M10W	M10W	M10W
	275			M10W	M10W	M10W	M10W	M10W	M10W

L

Calculation Formulas for Determining Angles, Distances and Radii

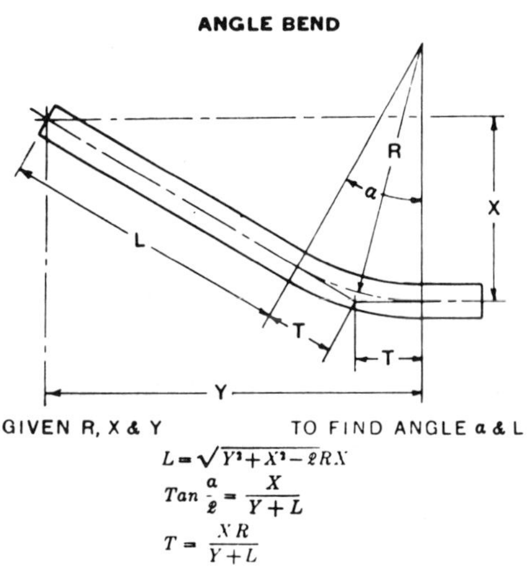

$$L = \sqrt{Y^2 + X^2 - 2RX}$$

$$Tan\,\frac{a}{2} = \frac{X}{Y + L}$$

$$T = \frac{XR}{Y + L}$$

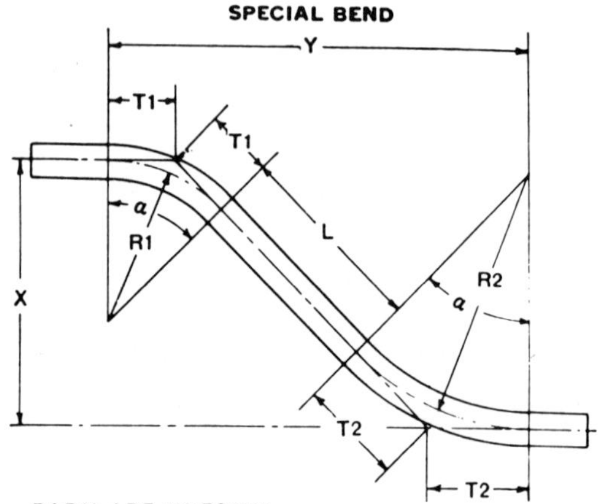

RADII ARE UNEQUAL

GIVEN R_1, R_2, X & Y TO FIND ANGLE a & L

WHERE $Y \geq \sqrt{2(R_1+R_2)X - X^2}$

$$L = \sqrt{Y^2 + X^2 - 2(R_1 + R_2)X}$$

$$Tan\,\frac{a}{2} = \frac{X}{Y + L}$$

$$T_1 = \frac{R_1 X}{Y + L}$$

$$T_2 = \frac{R_2 X}{Y + L}$$

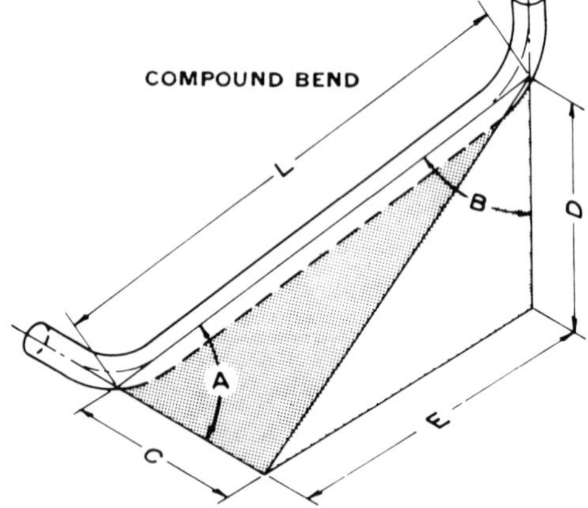

GIVEN C, D & E TO FIND ANGLE A, ANGLE B & L

$$L = \sqrt{C^2 + D^2 + E^2}$$

$$Cos\ A = \frac{C}{L}$$

$$Cos\ B = \frac{D}{L}$$

OFFSET BENDS

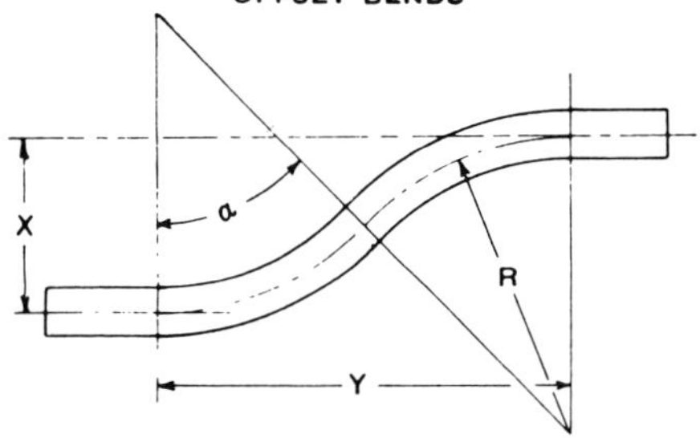

GIVEN R & X
TO FIND ANGLE α & Y

$$Y = \sqrt{4RX - X^2}$$

$$Tan\ \frac{\alpha}{2} = \frac{X}{Y}$$

GIVEN Y & X
TO FIND ANGLE α & R

$$R = \frac{Y^2 + X^2}{4X}$$

$$Tan\ \frac{\alpha}{2} = \frac{X}{Y}$$

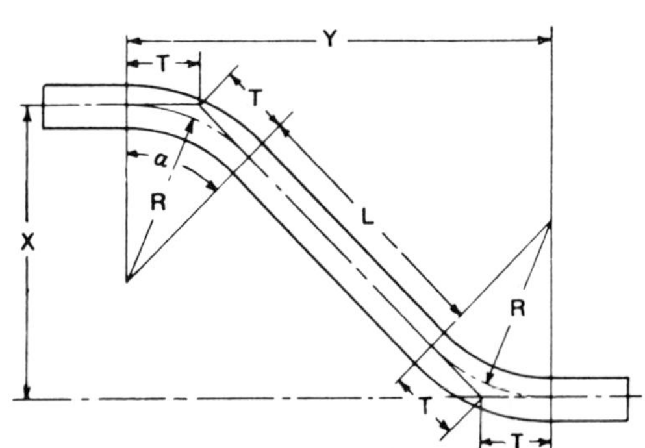

GIVEN R, X & Y TO FIND ANGLE α & L

WHERE $Y \geq \sqrt{4RX - X^2}$

$$L = \sqrt{Y^2 + X^2 - 4RX}$$

$$Tan\ \frac{\alpha}{2} = \frac{X}{Y + L}$$

$$T = \frac{RX}{Y + L}$$

M

Tables for Length of an Arc, Subtended by an Angle

Table 1. Any Angle

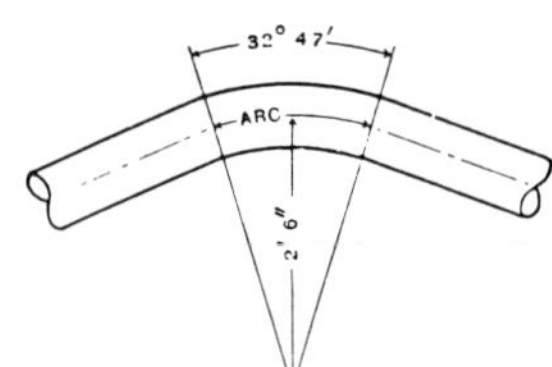

Example

To obtain length of arc 32° 47′–Radius 2′-6″ or 2.5′

Constant 32° = .5585

47′ = .01367

.57217 X 2.5′ = 1.43′ = 1′5 3/16″

For Bends over 90° Constants may be added to equal total angle

Length of arc in 120° Bend = 90° = 1.5708

30° = .5236

2.0944 X Radius

Tangent Ends (straight part of bend) must be added to arc length. For minimum length of Tangent Ends see opposite page.

Degrees–Radius Equal 1						Minutes–Radius Equal 1			
Degrees	Constant	Degrees	Constant	Degrees	Constant	Minutes	Constant	Minutes	Constant
1	.0175	31	.5411	61	1.0647	1	.00029	31	.00901
2	.0349	32	.5585	62	1.0821	2	.00058	32	.00930
3	.0524	33	.5760	63	1.0996	3	.00087	33	.00959
4	.0698	34	.5934	64	1.1170	4	.00116	34	.00989
5	.0873	35	.6109	65	1.1345	5	.00145	35	.01018
6	.1047	36	.6283	66	1.1519	6	.00174	36	.01047
7	.1222	37	.6458	67	1.1694	7	.00203	37	.01076
8	.1396	38	.6632	68	1.1868	8	.00233	38	.01105
9	.1571	39	.6807	69	1.2043	9	.00261	39	.01134
10	.1745	40	.6981	70	1.2217	10	.00290	40	.01163
11	.1920	41	.7156	71	1.2392	11	.00319	41	.01192
12	.2094	42	.7330	72	1.2566	12	.00349	42	.01221
13	.2269	43	.7505	73	1.2741	13	.00378	43	.01250
14	.2443	44	.7679	74	1.2915	14	.00407	44	.01279
15	.2618	45	.7854	75	1.3090	15	.00436	45	.01309
16	.2793	46	.8029	76	1.3265	16	.00465	46	.01338
17	.2967	47	.8203	77	1.3439	17	.00494	47	.01367
18	.3142	48	.8378	78	1.3614	18	.00523	48	.01396
19	.3316	49	.8552	79	1.3788	19	.00552	49	.01425
20	.3491	50	.8727	80	1.3963	20	.00581	50	.01454
21	.3665	51	.8901	81	1.4137	21	.00610	51	.01483
22	.3840	52	.9076	82	1.4312	22	.00639	52	.01512
23	.4014	53	.9250	83	1.4486	23	.00669	53	.01541
24	.4189	54	.9425	84	1.4661	24	.00698	54	.01570
25	.4363	55	.9599	85	1.4835	25	.00727	55	.01599
26	.4538	56	.9774	86	1.5010	26	.00756	56	.01628
27	.4712	57	.9948	87	1.5184	27	.00785	57	.01658
28	.4887	58	1.0123	88	1.5359	28	.00814	58	.01687
29	.5061	59	1.0297	89	1.5533	29	.00843	59	.01716
30	.5236	60	1.0472	90	1.5708	30	.00872	60	.01754

Table 2. 90° Angle—Radii 1″ to 15′-11″

LENGTH OF ARC SUBTENDED BY 90° ANGLE
RADIUS = 1″ TO 15′-11″-APPROXIMATED TO NEAREST 1/16″

Radius Ft.	0-inch	1-inch	2-inch	3-inch	4-inch	5-inch	6-inch	7-inch	8-inch	9-inch	10-inch	11-inch
0	---	1 9/16"	3 1/8"	4 3/4"	6 1/4"	7 7/8"	9 3/8"	11"	1'-0 9/16"	1'-2 1/8"	1'-3 11/16"	1'-5 1/4"
1	1'-6 7/8"	1'-8 7/16"	1'-10"	1'-11 9/16"	2'-1 1/8"	2'-2 11/16"	2'-4 1/4"	2'-5 7/8"	2'-7 7/16"	2'-9"	2'-10 9/16"	3'-0 1/8"
2	3'-1 11/16"	3'-3 1/4"	3'-4 13/16"	3'-6 3/8"	3'-8"	3'-9 9/16"	3'-11 1/8"	4'-0 11/16"	4'-2 1/4"	4'-3 13/16"	4'-5 7/16"	4'-7"
3	4'-8 9/16"	4'-10 1/8"	4'-11 11/16"	5'-1 1/4"	5'-2 13/16"	5'-4 3/8"	5'-6"	5'-7 9/16"	5'-9 1/8"	5'-10 11/16"	6'-0 1/4"	6'-1 13/16"
4	6'-3 3/8"	6'-5"	6'-6 9/16"	6'-8 1/8"	6'-9 11/16"	6'-11 1/4"	7'-0 13/16"	7'-2 3/8"	7'-4"	7'-5 9/16"	7'-7 1/8"	7'-8 11/16"
5	7'-10 1/4"	7'-11 13/16"	8'-1 3/8"	8'-3"	8'-4 9/16"	8'-6 1/8"	8'-7 11/16"	8'-9 1/4"	8'-10 13/16"	9'-0 3/8"	9'-1 15/16"	9'-3 9/16"
6	9'-5 1/8"	9'-6 5/8"	9'-8 1/4"	9'-9 13/16"	9'-11 3/8"	10'-0 15/16"	10'-2 1/2"	10'-4 1/8"	10'-5 11/16"	10'-7 1/4"	10'-8 13/16"	10'-10 3/8"
7	10'-11 15/16"	11'-1 1/2"	11'-3 1/8"	11'-4 5/8"	11'-6 1/4"	11'-7 13/16"	11'-9 3/8"	11'-10 15/16"	12'-0 1/2"	12'-2 1/16"	12'-3 5/8"	12'-5 1/4"
8	12'-6 13/16"	12'-8 3/8"	12'-9 15/16"	12'-11 1/2"	13'-1 1/16"	13'-2 5/8"	13'-4 1/4"	13'-5 3/4"	13'-7 3/8"	13'-8 15/16"	13'-10 1/2"	14'-0 1/16"
9	14'-1 5/8"	14'-3 1/4"	14'-4 3/4"	14'-6 3/8"	14'-7 15/16"	14'-9 1/2"	14'-11 1/16"	15'-0 5/8"	15'-2 1/4"	15'-3 3/4"	15'-5 3/8"	15'-6 7/8"
10	15'-8 1/2"	15'-10 1/16"	15'-11 5/8"	16'-1 1/4"	16'-2 3/4"	16'-4 3/8"	16'-5 7/8"	16'-7 1/2"	16'-9 1/16"	16'-10 5/8"	17'-0 3/16"	17'-1 3/4"
11	17'-3 3/8"	17'-4 7/8"	17'-6 1/2"	17'-8 1/16"	17'-9 5/8"	17'-11 3/16"	18'-0 3/4"	18'-2 3/8"	18'-3 7/8"	18'-5 1/2"	18'-7 1/16"	18'-8 5/8"
12	18'-10 3/16"	18'-11 3/4"	19 - 1 5/16"	19'-2 7/8"	19'-4 1/2"	19'-6 1/16"	19'-7 5/8"	19'-9 3/16"	19'-10 3/4"	20'-0 5/16"	20'-1 7/8"	20'-3 1/2"
13	20'-5 1/16"	20'-6 5/8"	20'-8 3/16"	20'-9 3/4"	20'-11 5/16"	21'-0 7/8"	21'-2 1/2"	21'-4 1/16"	21'-5 5/8"	21'-7 3/16"	21'-8 3/4"	21'-10 5/16"
14	21'-11 7/8"	22'-1 1/2"	22'-3"	22'-4 5/8"	22'-6 3/16"	22'-7 3/4"	22'-9 5/16"	22'-10 7/8"	23'-0 1/2"	23'-2"	23'-3 5/8"	23'-5 3/16"
15	23'-6 3/4"	23'-8 1/4"	23'-9 7/8"	23'-11 1/2"	24'-1"	24'-2 5/8"	24'-4 3/16"	24'-5 3/4"	24'-7 5/16"	24'-8 13/16"	24'-10 7/16"	25'-0"

Tangent Ends (straight part of bend) must be added to above arc lengths. For minimum Tangent Ends see opposite page.

N

Updates for Second Edition

In the 10 years since this book was first published, a number of changes have taken place among bending companies. Some companies have changed their names and ownerships, some have gone out of business, and several new companies have emerged.

Machines have been improved and become even more expensive, and today more than 40 companies represent cold bending in the United States alone. Yet, the bending business itself has changed very little. Regrettably, there are still no national standards for bending, there are no trade qualifications for operators, and there is no standardization of machinery.

Induction bending has seen the withdrawal of one manufacturer, Gregson, in England, and the emergence of a new one, Schafer, in Germany. Wilhelm Schafer Maschinenbau GmbH & Co. was already well-established with a variety of products related to pipe before introducing its first large induction bending machines in 1986. Unlike Cojafex, which has no in-house capability, Schafer has a separate division dealing exclusively with job shop induction bending.

However, neither company, nor Dai-Ichi, the principal Japanese induction bending company with more machines in their shops than there are in the whole of North America, has made any real attempt to tackle the small bore piping market, which accounts for about 80 percent of all piping, 95 percent of which uses standard, butt-weld elbows for changes of direction. All three manufacturers have machines that can make 1½xD bends with wall-thinning tolerances within those specified for a standard fitting, but they focus on the larger pipe sizes from about 4 inches upwards.

One English company recently developed a small, portable induction bender that could produce 1½xD bends in 4-inch pipe, but the company had no real knowledge of what was required for pipe spool production and has a long way to go before producing a marketable unit. Similarly, a U.S. company has also developed a small, portable machine but has never pursued the research and development needed to produce acceptable quality 1½xD bends.

Another problem peculiar to induction bending is that of quality and cost. Because heat is involved, many jobs require a preproduction test bend for each heat of material involved. Thus, for a job involving a number of bends which must be made in pipe from, for example, three different heats, a preproduction bend must be made for each heat, and samples of the material must be taken from the tangent areas, the transition zones, and at least three bend areas (see Figure N-1). All the test coupons are examined in detail by the metallurgist, who determines the correct bending temperature and speed and any postbend heat treatment that may also be required.

Naturally, this adds considerably to the final cost of the job. Yet, most induction bending companies maintain extensive records of the

Figure N-1. (Courtesy of S.A. Fabricom N.V. Belgium.)

jobs they have completed and consider this requirement unnecessary. Only if they are dealing with a new type of material or have no prior records detailing the job would they consider it necessary to carry out a preproduction test bend and extract all the coupons required for full metallurgical examination.

It would seem, therefore, the industry is still a long way from undertaking more cost-effective methods of bending pipe and fabricating it. The technology exists that can put a pipe end into perfect round to a tolerance of ±.001 inch or weld pipe ends together in less than one second without the use of any weld material. The capability also exists to produce a family of small, portable induction benders that could cover almost all the small-bore pipe sizes up to and including 6 inches for making bends in the 1½xD to 5xD range.

The potential cost savings could be staggering, especially if engineers were to design piping systems with slightly larger radii than the conventional 1½xD. This savings has already been proven by a study recently conducted by a large power company. It determined that as much as 95 percent of all corners in piping from ½ inch through 4 inches could be bends rather than welded elbows and that, in most cases, the radii could be as large as 5xD. The net cost savings for one power plant alone was a staggering $17,800,000.

Clearly, the industry has a long way to go before it becomes truly cost-effective, and perhaps this is best illustrated in the area of computer software which is initially focusing on the tube bending and fabricating section of the industry.

Until recently, the use of computer software for automating the bending processes has been slow to gain acceptance, although, over the years, software has been developed for specific areas, mainly in respect to the bending machines themselves. Examples of this can be seen in Chapter 2 of this book, which illustrates computer usage to determine optimum conditions of bend to satisfy a customer's needs. Cold bending also uses computers and associated software, but here, too, such software is related to a specific manufacturer's machine and deals principally with the actual bending process.

For routing of tube and pipe, add-on software packages are available from most of the computer-aided design (CAD) software companies. The complexity of these packages can range from very simple 2-D drawings to complex systems that provide dynamic 3-D rendered examples.

Software is also available to handle the calculations needed for cold bending. Three such packages that illustrate the diverse presentation format available with today's hardware are HP5M from MC2 Engineering Software, which is DOS-based; Tube-CAD from Adaptive Motion Control Systems, which works with AutoCAD®; and the USI-Bend Calculator from Unlimited Solutions, which runs under Microsoft Windows®. Most of the calculations automated by these software packages are described in this book. The first two packages can also draw a 3-D wire representation of the tube and download the information to a CAD system.

Some companies have used spreadsheets or generic costing software systems to help automate the fabrication process. These systems have been developed by some of the larger tubing job shops and are not integrated with any other software systems; often, they are difficult to maintain.

One software for tube design that is easier to use is the Tube Expert™ from Unlimited Solutions.This is a complete product configuration package optimized for the tubing industry. One of the many areas it addresses is the costing of the complete tube fabrication job. It has a graphical user interface (see Figure N-2) where information is entered

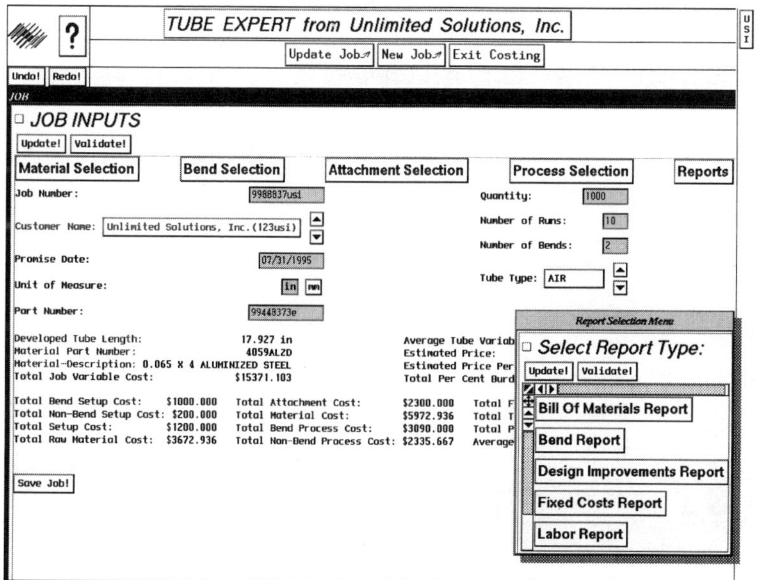

Figure N-2. Tube Expert™ graphical user interface.

into the system and proper information is selected from a menu. In addition to the intuitive interface, complete contextual help guides less-experienced users through the design analysis process (see Figure N-3).

All information concerning materials, attachments, processes, machines, tooling, setup time, process time, and scrap can easily be entered by the user into the database.

All output information is also stored in the same database for developing quotes, bills of material, process plans, and reports and for use by other company computer systems such as accounting, material requirements planning (MRP), and inventory control.

Software such as this can help enormously to automate the costing process by providing, among other capabilities:

1. Accurate costs in minutes instead of hours.
2. Consistent costs without guesswork or omissions.
3. Performance of "what if" scenarios to allow on-the-fly design changes.

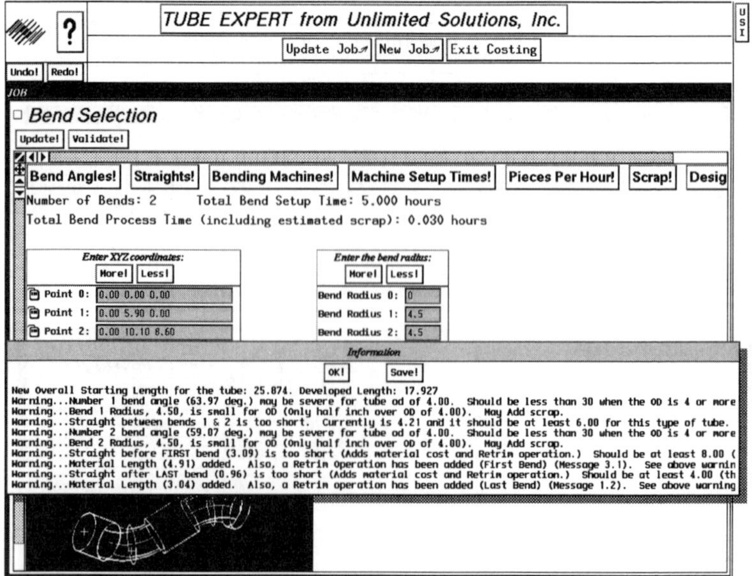

Figure N-3. Contextual help on Tube Expert™.

4. Creation of custom cost reports and analyses.
5. Creation of bills of material and process plans.
6. Development of complete process costs including setup, process time, and scrap.

All too often, crucial elements are omitted during the costing process because of oversight and lumping every movement of material and associated labor cost into "shop overhead." Particularly among job shops, true costs are rarely fully established before a job is accepted, which often causes perplexity when the job is shown to have lost money instead of making a profit.

It is long past time for the industry to enter the computer age and become more cost efficient.

Index